MODERNIZING CRIME STATISTICS

Report 2–New Systems for Measuring Crime

Panel on Modernizing the Nation's Crime Statistics

Janet L. Lauritsen and Daniel L. Cork, *Editors*

Committee on National Statistics
Committee on Law and Justice

Division of Behavioral and Social Sciences and Education

A Consensus Study Report of
The National Academies of
SCIENCES · ENGINEERING · MEDICINE

THE NATIONAL ACADEMIES PRESS
Washington, DC
www.nap.edu

THE NATIONAL ACADEMIES PRESS 500 Fifth Street, NW Washington, DC 20001

The project that is the subject of this report was supported by the Bureau of Justice Statistics, Office of Justice Programs, and the Federal Bureau of Investigation, via a grant from the National Science Foundation (No. SES-1024012) that permits a consortium of federal agencies to support the work of the Committee on National Statistics. Any opinions, findings, conclusions, or recommendations expressed in this publication do not necessarily reflect the views of the organizations or agencies that provided support for the project.

International Standard Book Number-13: 978-0-309-47261-6
International Standard Book Number-10: 0-309-47261-X
Digital Object Identifier: https://doi.org/10.17226/25035

Additional copies of this publication are available from the National Academies Press, 500 Fifth Street, NW, Keck 360, Washington, DC 20001; (800) 624-6242 or (202) 334-3313; http://www.nap.edu.

Printed in the United States of America

Suggested citation: National Academies of Sciences, Engineering, and Medicine. (2018). *Modernizing Crime Statistics—Report 2: New Systems for Measuring Crime.* Washington, DC: The National Academies Press. https://doi.org/10.17226/25035.

The National Academies of

SCIENCES • ENGINEERING • MEDICINE

The **National Academy of Sciences** was established in 1863 by an Act of Congress, signed by President Lincoln, as a private, nongovernmental institution to advise the nation on issues related to science and technology. Members are elected by their peers for outstanding contributions to research. Dr. Marcia McNutt is president.

The **National Academy of Engineering** was established in 1964 under the charter of the National Academy of Sciences to bring the practices of engineering to advising the nation. Members are elected by their peers for extraordinary contributions to engineering. Dr. C. D. Mote, Jr., is president.

The **National Academy of Medicine** (formerly the Institute of Medicine) was established in 1970 under the charter of the National Academy of Sciences to advise the nation on medical and health issues. Members are elected by their peers for distinguished contributions to medicine and health. Dr. Victor J. Dzau is president.

The three Academies work together as the **National Academies of Sciences, Engineering, and Medicine** to provide independent, objective analysis and advice to the nation and conduct other activities to solve complex problems and inform public policy decisions. The National Academies also encourage education and research, recognize outstanding contributions to knowledge, and increase public understanding in matters of science, engineering, and medicine.

Learn more about the National Academies of Sciences, Engineering, and Medicine at www.nationalacademies.org.

The National Academies of
SCIENCES • ENGINEERING • MEDICINE

Consensus Study Reports published by the National Academies of Sciences, Engineering, and Medicine document the evidence-based consensus on the study's statement of task by an authoring committee of experts. Reports typically include findings, conclusions, and recommendations based on information gathered by the committee and the committee's deliberations. Each report has been subjected to a rigorous and independent peer-review process and it represents the position of the National Academies on the statement of task.

Proceedings published by the National Academies of Sciences, Engineering, and Medicine chronicle the presentations and discussions at a workshop, symposium, or other event convened by the National Academies. The statements and opinions contained in proceedings are those of the participants and are not endorsed by other participants, the planning committee, or the National Academies.

For information about other products and activities of the National Academies, please visit nationalacademies.org/whatwedo.

PANEL ON MODERNIZING THE NATION'S CRIME STATISTICS

JANET L. LAURITSEN,* (*Chair*), Department of Criminology and Criminal Justice, University of Missouri–St. Louis

DANIEL B. BIBEL, Crime Reporting Unit, Massachusetts State Police, Maynard (retired)

JONATHAN P. CAULKINS, Heinz College of Information Systems and Public Policy, Carnegie Mellon University

KIM ENGLISH, Division of Criminal Justice, Colorado Department of Public Safety

ROBERT M. GOERGE, Chapin Hall at the University of Chicago

NOLA M. JOYCE, Philadelphia Police Department (retired)

DAVID MCDOWALL, Violence Research Group, University at Albany, State University of New York

JENNIFER H. MADANS, National Center for Health Statistics

MICHAEL D. MALTZ, Department of Criminology, Law, and Justice, University of Illinois at Chicago (emeritus) and Criminal Justice Research Center, Ohio State University

MICHAEL C. MILLER, Coral Gables Police Department, Florida

JAMES J. NOLAN, III, Department of Sociology and Anthropology, West Virginia University

AMY O'HARA, Stanford Institute for Economic Policy Research, Stanford University

JOHN V. PEPPER, Department of Economics, University of Virginia

ALEX R. PIQUERO, School of Economic, Political, and Policy Sciences, University of Texas at Dallas

JEFFREY L. SEDGWICK,† Justice Research and Statistics Association, Washington, DC

JAMES P. LYNCH, Department of Criminology and Criminal Justice, University of Maryland (*Consultant to the panel*)

PAUL K. WORMELI, Wormeli Consulting, LLC, Ashburn, Virginia (*Consultant to the panel and liaison from Committee on Law and Justice*)

DANIEL L. CORK, *Study Director*

SETH HAUSER, *Senior Program Officer (February-October 2015)*

EDWARD SPAR, *Senior Program Officer (through September 2014)*

MICHAEL SIRI, *Program Associate*

JORDYN WHITE, *Program Officer*

---

*Served as member of the panel from formation through December 8, 2014.

†Served as chair of the panel from formation through December 8, 2014, following appointment as executive director of the Justice Research and Statistics Association.

COMMITTEE ON NATIONAL STATISTICS
2018

COMMITTEE ON LAW AND JUSTICE
2018

ROBERT D. CRUTCHFIELD (*Chair*), Department of Sociology, University of Washington (emeritus)
SALLY S. SIMPSON (*Vice-chair*), Department of Criminology and Criminal Justice, University of Maryland
PREETI CHAUHAN, John Jay College of Criminal Justice
KIMBERLÉ W. CRENSHAW, School of Law, University of California, Los Angeles
JOHN J. DONOHUE III, Stanford Law School, Stanford University
MARK S. JOHNSON, Department of Community and Family Medicine, Howard University
MARK A.R. KLEIMAN, Marron Institute of Urban Management, New York University
JAMES P. LYNCH, Department of Criminology and Criminal Justice, University of Maryland
KAREN MATHIS, Institute for the Advancement of the American Legal System, University of Denver
DANIEL S. NAGIN, H. John Heinz III College of Public Policy and Information Systems, Carnegie Mellon University
ANNE MORRISON PIEHL, Department of Economics and Program in Criminal Justice, Rutgers University
STEVEN RAPHAEL, Goldman School of Public Policy, University of California, Berkeley
LAURIE O. ROBINSON, Department of Criminology, Law, and Society, George Mason University
CYNTHIA RUDIN, Departments of Computer Science and Electrical and Computer Engineering and Prediction Analysis Lab, Duke University
SUSAN B. SORENSON, School of Social Policy and Practice, University of Pennsylvania
LINDA A. TEPLIN, Feinberg School of Medicine, Northwestern University
HEATHER ANN THOMPSON, Department of History, University of Michigan
BRUCE WESTERN, Department of Sociology, Harvard University

KATHI GRASSO, *Director*

Acknowledgments

THE PANEL ON MODERNIZING THE NATION'S CRIME STATISTICS is pleased to submit this second and final report on modernizing the collection of data on crime in the United States. Of necessity, because most of the data gathering for this final report took place during preparation of our first report, these acknowledgments are virtually identical to those in Report 1. We trust that it will be understood that the repetition is not meant to diminish our appreciation of the time and talent of all those who have contributed to the panel's work.

Both of our sponsoring agencies—the Bureau of Justice Statistics (BJS) and the Criminal Justice Information Services (CJIS) Division of the Federal Bureau of Investigation (FBI)—committed considerable resources and attention to our study, for which we are greatly appreciative. Former BJS director William Sabol, deputy director Gerard Ramker, and deputy and acting director Jeri Mulrow provided constant support; we owe much to senior statistical advisor Allen Beck for his discussions of his participation with the United Nations Office on Drugs and Crime (UNODC) expert work group on crime classification; and we thank the other BJS staff who gave unyieldingly of their time, including Howard Snyder, Michael Planty, and Erica Smith. At FBI/CJIS, Amy Blasher provided valuable assistance and liaison to our panel as Uniform Crime Reporting (UCR) Program manager, Crime Statistics Management unit chief, and UCR Modernization manager; Samuel Berhanu provided updates to the panel after succeeding Blasher as Crime Statistics Management unit chief. Michelle Klimt was instrumental in the creation of this project during her time as chief of the CJIS Law Enforcement Support Section, and John Derbas has continued to provide assistance. We have also benefited from extensive interaction with statisticians James Noonan and Cynthia Barnett-Ryan, from the engagement of Randall Thysse in our work, and from a discussion with executive assistant director for the Science and Technology Branch Amy Hess during her tenure in that position. The U.S. Office of Management and Budget provided the essential encouragement for BJS and FBI to jointly sponsor this review; we are particularly grateful to Rochelle Martinez and the Statistical and Science Policy Office.

The production of a modern classification of crime, in Report 1, is arguably the major product and weightiest conceptual lift of this project. In the development of our proposed classification—from which stems the discussion in this second report—we remain greatly indebted to the work of a UNODC expert work group established to draft an International Classification of Crime for Statistical Purposes. Angela Me, chief of the UNODC's Research and Trend Analysis Branch, traveled to speak to our panel in person at our second meeting, and Michael Jandl (research officer in the Statistics and Survey Division) graciously stepped in on short notice to give a brief overview of the UNODC work by teleconference at our first meeting. We note that the UNODC work group is continuing to work out the methodological and implementation issues of the kind we address in this report and look forward to continued cooperation, and have appreciated the opportunity to discuss the work at conferences on international crime, justice, and governance statistics.

With continued appreciation, we refer readers to Appendix B of Report 1—the detailed list of participant stakeholders from our panel's data-gathering meetings. We again thank all the speakers and participants who enriched our discussions with their wide-ranging perspectives (and shared interest in high-quality crime and justice data), and look forward to continued collaboration in implementation stages.

At the National Academies of Sciences, Engineering, and Medicine, Michael Siri provided expert and always considerate assistance in managing the logistics of this very active study panel. For a little more than 6 critical months in 2015, the panel had the invaluable assistance of Seth Hauser, who took to a short-term assignment as senior program officer with gusto and deftly constructed the agenda for two of our panel meetings. Program officer Jordyn White assisted with our proceedings. Daniel Cork served as study director for this project and deserves special recognition. His skills for incorporating large amounts of technical and organizational data on crime statistics with insights from data users, providers, and panel members were inimitable and invaluable to the writing of this report.

This Consensus Study Report was reviewed in draft form by individuals chosen for their diverse perspectives and technical expertise. The purpose of this independent review is to provide candid and critical comments that will assist the National Academies of Sciences, Engineering, and Medicine in making each published report as sound as possible and to ensure that it meets the institutional standards for quality, objectivity, evidence, and responsiveness to the study charge. The review comments and draft manuscript remain confidential to protect the integrity of the deliberative process.

We thank the following individuals for their review of this report: Thomas Abt, Harvard Law School; John J. Donohue III, Stanford Law School; John Jackson, Patrol Region 3, Houston Police, Houston, TX; Angela Me, Research and Trend Analysis Branch, United Nations Office on Drugs and Crime; Peter Reuter, School of Public Policy and Department of Criminology, University of Maryland, College Park; John Roman, NORC at the University of Chicago; and Max Schlueter, The Crime Research Group, Montpelier, VT.

Although the reviewers listed above provided many constructive comments and suggestions, they were not asked to endorse the conclusions or recommendations of this report nor did they see the final draft before its release. The review of this report was overseen by Alfred Blumstein, H. John Heinz III College of Public Policy and Information Systems, Carnegie Mellon University, and Philip J. Cook, Sanford School of Public Policy, Duke University. They were responsible for making certain that an independent examination of this report was carried out in accordance with the standards of the National Academies and that all review comments were carefully considered. Responsibility for the final content rests entirely with the authoring committee and the National Academies.

Janet L. Lauritsen, *Chair*
Panel on Modernizing the Nation's Crime Statistics

Contents

List of Figures, Tables, and Boxes

FIGURE

TABLE

BOXES

Summary

AS A NATION, we know much less about crime than we should:

- Word of new thefts of data from businesses or government agencies seems to arrive near daily—thefts that put thousands, if not millions, of Americans at risk of identity theft and serious financial harm. Cyberspace is not only an active and ongoing international battlefront, but also a perpetual domestic crime scene. The nation's crime statistics have not been well equipped to measure cybercrimes such as the deployment of malware or coordinated attacks on computer networks.
- Drug production, sale, and possession are important crime categories, as is highlighted by today's high rates of opioid overdose. But levels of these activities are not well characterized in current crime statistics; counts of seizures and arrests for drug law violations may say as much about law enforcement's reaction to illegal drug activity than the levels and changes in the drug activity itself. Since 1991 the Office of National Drug Control Policy has periodically published estimates of the quantity (in metric tons) and dollar value of the major drug markets, along with estimates of the numbers of people using these drugs with various intensities. Those are arguably more informative metrics for characterizing the extent of drug-law violating than are arrests for those violations, yet they have not historically been part of conventional reporting on crime in the United States.
- Every day, inventory and resources "disappear" from the stocks of the nation's retailers and businesses, some for explainable reasons but much through criminal acts such as theft/pilferage by employees or fraudulent resale. Collectively, the phenomenon is known as shrinkage and is poorly understood by the public, in part because businesses do not commonly

report all of their losses to law enforcement officials even though the billions of dollars in losses almost inevitably affect all consumers.

- Current national crime statistics lack the detail and the timeliness to address important concerns, even about offenses that are commonly thought to be well measured. For instance, in 2006 and 2015 in particular, news accounts and general perception fueled concerns about major increases in homicide despite the fact that the current homicide rate is less than half the rate of the 1980s and 1990s. Some local law enforcement agencies were equipped to use their own data resources to characterize the local problems, and databases created by media outlets and advocacy or other groups provided some additional insight. But the nation as a whole was hindered by the typical 10-month gap between the end of the calendar year and the release of current national crime statistics, unable to understand whether local patterns were part of broader regional or national patterns or whether increased homicide activities were limited to specific forms of homicide such as drug-related murder.

These are some examples of knowledge gaps in the nation's current crime statistics. This panel's Report 1 on *Modernizing Crime Statistics* (National Academies of Sciences, Engineering, and Medicine, 2016) focused on the definition and concept of crime; this final report considers how the nation can remedy these important knowledge gaps.

CURRENT NATIONAL CRIME STATISTICS, AND THIS PANEL'S WORK

For decades following its establishment in 1930, the Summary Reporting System of the Uniform Crime Reporting (UCR) Program (SRS, or UCR Summary) was the principal source of U.S. crime statistics, compiled and published by the Federal Bureau of Investigation (FBI)—since 1959, in a report bearing the prominent title *Crime in the United States.* The UCR data are aggregated tallies of offenses reported to or otherwise known to law enforcement agencies; the source data are voluntarily contributed to the national collection by way of a network of state-level coordinating bodies. Beginning in 1972, the nation's crime statistics were joined by an essential, complementary data collection—what is now known as the National Crime Victimization Survey (NCVS) sponsored by the Bureau of Justice Statistics (BJS). The NCVS uses household interviews to produce estimates of criminal victimizations that are *not* necessarily reported to law enforcement, the so-called "dark figure" of crime. More recently, a mid-1980s evaluation led the FBI to institute the National Incident-Based Reporting System (NIBRS)—characterized as the future of UCR data collection but marred by decades of slower-than-anticipated adoption by state and local law enforcement agencies.

In 2016, the FBI formalized a decision to retire the UCR Summary system by 2021 in favor of a full-fledged NIBRS.

At the suggestion of the U.S. Office of Management and Budget (OMB), this Panel on Modernizing the Nation's Crime Statistics was engaged to review the whole U.S. crime statistics enterprise. Specifically, the panel was tasked with a three-part charge by BJS and FBI, the full text of which is in Appendix A. Report 1 covered the first, *substantive* plank in that charge: performing a comprehensive reassessment of what is meant by crime in U.S. crime statistics, and recommending a new classification of crime to organize measurement efforts. This report covers the two remaining planks of the charge:

- *Methodological* issues, assessing the adequacy and appropriateness of current crime statistics collections and suggesting the best conceptual means of collecting data based on the recommended classification, including the integration of data on specific crime types from non-BJS or FBI sources and enabling the identification and measurement of emerging crime types going forward; and
- *Implementation* issues, describing how national crime data collection should proceed, practically and effectively, including suggesting how to leverage available information technology assets, accommodate the demands of crime data stakeholders, and moderate the burden placed on crime data providers while being mindful of the voluntary nature of crime data reporting.

By the nature of panel studies like this one, and given the sheer extent of the panel's charge, this report is necessarily a conceptual blueprint for modernizing crime statistics rather than a precise and detailed implementation plan. But it serves to highlight the importance of the work to better understanding key national concerns.

REDEFINING AND MEASURING CRIME

Pursuant to our charge, we take a broad perspective on crime by defining it generally as socially unacceptable behavior that causes harm or threatens harm against another. Our behavioral definitions are not tethered to the specific legal language of federal or state criminal codes, and so treat as "crime" some things that might commonly be construed as tortious or civil because they are usually adjudicated in civil courts or regulatory proceedings. By this, and by construction of our proposed classification of crime offenses for statistical purposes, we do not suggest that any single offense or group of offenses is more important than another, just that all are essential for a complete, composite picture of crime.

This broader definition of crime is different from decades of crime statistics practice, and a brief trace of historical themes puts some fundamental issues in stark relief. Following a short-lived initial period of genuine enthusiasm at what could be learned from national crime statistics, the content of those statistics and the routines for producing them ossified—became overly rigid and stagnant—during the decades in which the UCR Summary system was the sole source, with major consequences. Initial candor about the completeness of the UCR data and clarity about their limitation to offenses reported to law enforcement became overwhelmed by an undue sense of comprehensiveness. Once the annual report of UCR data acquired its still-current name in 1959, *Crime in the United States* came to define "crime in the United States" in many eyes, restricting attention to a narrow set of specific offenses and further still on those offenses known to law enforcement. More fundamentally, the mindset of estimating crime—measuring its levels, trends, and characteristics—gave way to one of appearing to completely enumerate it, reducing the study of crime to the reporting of short-term changes in a few, broad categories of crime rates. The emergence of the NCVS as a counterweight to police-report-only data, and the introduction of NIBRS two decades later, made it possible to further widen the view, but the historical restrictions remain strong.

To provide detail on the coverage gaps in current crime statistics, this report matches the crime classification described in Report 1 to the coverage of current nationally compiled crime statistics sources (NIBRS and NCVS), as detailed in Appendix C and summarized in stylized form in Box S.1. Categories 1–5—which loosely correspond to the "traditional" violent interpersonal crimes and property offenses that dominate current crime statistics—are reasonably well covered in current crime data, albeit with some major conceptual differences. But offenses in our categories 6–11 are almost completely new terrain for U.S. crime statistics. Some (such as cybercrime) are "new" and emerging types of crime; others (such as environmental crimes and many types of fraud) would be new only in statistical tallies, having received little to no standardized national collection to date even though they have long been serious problems. A central observation is that any report that wishes to claim to present a comprehensive view of crime in the United States must pay serious attention to categories 6–11, as is done for categories 1–5.

One of our conclusions is meant as a strong corrective to decades of crime statistics being characterized or interpreted as an enumeration of all crime: A system of modern crime statistics will be more useful if it can provide more information and more flexibility for detailed analysis than simple summary counts.

> **Conclusion 2.1:** The aim of modern crime statistics is the effective measurement and estimation of crime. Accurate counting of offenses and incidents is important, but the nation's crime statistics will remain inadequate unless they expand to include more than just simple tallies with no associated measure of

Box S.1 "Traditional" and "New" Crime Types, In Proposed Classification

"Traditional" Crime
(generally covered by current statistics)

1 Acts leading to death or intending to cause death (*Examples:* murder, manslaughter/nonintentional homicide)
2 Acts causing harm or intending to cause harm to the person (*Examples:* assault, threat, extortion/blackmail, trafficking in persons, dangerous acts, stalking, harassment)
3 Injurious acts of a sexual nature (*Examples:* rape, sexual assault, sexual exploitation)
4 Acts of violence or threatened violence against a person that involve property (*Examples:* robbery, terroristic threats to infrastructure)
5 Acts against property only (*Examples:* burglary, theft, property damage (arson/burning), hacking and other acts against computer systems)

"New" Crime
(generally not well covered by current statistics)

6 Acts involving controlled substances (*Examples:* unlawful possession/use, cultivation/production, or trafficking of drugs)
7 Acts involving fraud, deception, or corruption (*Examples:* fraud, forgery/counterfeiting)
8 Acts against public order and authority (*Examples:* violations of gambling, alcohol, or tobacco regulations; acts against public order sexual standards; acts related to migration; acts against justice system)
9 Acts against public safety and national security (*Examples:* terrorism, organized crime offenses, trafficking of weapons/explosives)
10 Acts against the natural environment or against animals (*Examples:* environmental offenses, animal cruelty)
11 Other criminal acts not elsewhere (*Examples:* acts against military or tribal law, torture, war crimes)

uncertainty or capacity for disaggregation. Through the collection of associated attribute data, the crime statistics we suggest should—at minimum—enable the analysis of data in proper geographic, demographic, sociological, and economic context, and provide the raw material for important measures related to an offense (such as the harm it causes) in addition to its count.

In this study, the brand of crime data collection that we propose is incident-based, recording attributes describing incident, offender, and victim characteristics that are essential for disaggregation and analysis. For any particular offense type, the motivating questions should be what does the nation need to know about this kind of crime, and what is the most accurate or most feasible way to collect that information, with the desired degree of analytical detail. Uniformity of definitions and underlying concepts is essential, but absolute adherence to any one mode of data collection (e.g., police-report data versus survey-based data) is not; nor is observing any one portion of processing in the criminal justice system an imperative (i.e., some offenses may best or only be measured at the point of arrest or of charges being filed in court rather than earlier in the investigative phase).

Conclusion 2.1 is phrased as it is to convey that different measurement types may be needed for different specific offenses—and there are offense types for which counts alone may be inappropriate or misleading. This can be said of offenses falling on either side of our traditional/new divide, but is particularly acute for some of the new crime types:

- *Deployment of ransomware* involves malicious software that denies users access to their own data and computer systems until a ransom is paid. But there are material differences between a ransomware deployment that locks up an individual home-computer-user's machine and one that hobbles use of the information systems of entire hospital/medical groups or law enforcement departments; an offense count of 1 does little to convey what actually happened and improve understanding.
- Likewise, counts of *environmental violations* may be deceiving. Improperly disposing of a laptop computer battery and spilling a large amount of toxic materials into the water or air might each count as one event. Conversely, two events of the same size might have quite different consequences: Spilling the same number of barrels of oil into a water supply may not be equivalent to spilling the same number of barrels of oil in a nature reserve or an industrial site.

A related challenge in crime measurement is the timeliness of estimates, and the tradeoff between timeliness and precision. For some offense types including homicide, it may be essential to generate more frequent, interim estimates at the declared expense of increased uncertainty in addition to more refined final estimates.

STRUCTURE OF NEW NATIONAL CRIME STATISTICS COLLECTION

A related conclusion recognizes that no single data source can span the full range of crime; rather, a new crime statistics system will necessarily be a system of multiple data collection efforts:

> **Conclusion 2.3:** Improvement in the nation's crime statistics will require enhancements to and expansions of the current data collections, as well as new data collection systems for the historically neglected crime types highlighted by the proposed crime classification.

It follows from this conclusion that there is an important role going forward for the core methodologies currently at work in national crime statistics, albeit with refinements to those methodologies, as well as new approaches.

Data on offenses reported to or otherwise known to law enforcement—such as those embodied in NIBRS—will continue to be essential for U.S. crime statistics. Both of the above conclusions show strong and enthusiastic support for both the retirement of UCR Summary reporting and for full implementation of the National Crime Statistics Exchange (NCS-X) initiative (managed by BJS including funding from the FBI). The NCS-X initiative is working to target NIBRS development in a sample of jurisdictions so that the combined NIBRS-format data will provide a nationally representative sample, something that has not happened given the system's slow adoption rate for the past 30 years. But we emphasize a point made in closing Report 1: A full-local-compliance NIBRS is necessary but not sufficient. This is because a full NIBRS would still fall short of measuring all the offense types for which police-report data might be the most appropriate and accurate avenue, and it covers some offenses (notably assaults and threats, as well as the controlled substance offenses in our category 6) in a relatively uninformative way.

Another important point on the future of voluntary-contribution police-report crime data is that they will succeed and endure to the extent that they become a largely invisible by-product of daily operations at local law enforcement agencies. Individual law enforcement agencies differ in the priority they assign to the preparation of incident reports by officers and the tools available to produce them (ranging from hard-copy paper forms to mobile computer terminals in patrol cars). But, to the extent that more agencies see reporting (and the resulting crime statistics) as important means of both doing their work and of bolstering their accountability to the communities they serve, the more national crime statistics and their users will be able to benefit from a system where reporting to national collections is not a chore or something done out of a sense of obligation.

Self-report survey information on criminal victimization—including but not limited to the NCVS—will also remain a pillar of U.S. national crime statistics. Survey data on crime are necessarily limited in the degree to which the resulting data can be disaggregated along useful analytical lines; certain crime types are still, relatively speaking, rare events in the population at large, so the survey's sample size would have to be unworkably large to support fine-grained estimates. But survey data should remain part of the crime statistics infrastructure because of their unique, conceptual strength—their ability to estimate the level of victimization that is not reported to law enforcement authorities. In this report and this project, we say less about the NCVS and survey measures than about police-report methodologies; this is due to the fact that the NCVS has been in a state of near-continuous refinement and improvement, rather than a reflection on the importance of the data.

The frontiers for methodological and implementation work on the survey information side of national crime statistics are fairly well known and already active. These new frontiers include finding ways to directly or indirectly generate some subnational estimates from the data and developing techniques for measuring crime types (notably rape and sexual assault) that are particularly sensitive and difficult to measure, as well as making more effective use of interview time (including with households with no victimization experiences to report) to generate the kind of community-context variables that might be used in drawing meaningful comparisons between crime levels in multiple jurisdictions. It must also be noted that there are some offense types for which survey-based measures of perceived victimization, even at the expense of geographic precision, may actually be the most accurate available, as well as the most salient to data users and stakeholders. Identity theft is one crime type where NCVS has already proven an effective measure, and others might include other classes of fraud as well as stalking and harassment offenses. Again, though, the quality of victimization survey data is contingent on eliciting sufficient responses on possible crime incidents in order to draw inferences, and so it is critical that the NCVS be adequately resourced to fulfill its purposes.

To these two core methodologies, we add a third: the aggregation of information through a crime measurement clearinghouse function, perhaps primarily administrative record-type data drawn from external resources including federal and state agencies. These new types of information are necessary for the new types of crime in our classification, and developing them will require some substantial initial attention and investment. This is merely because they have not been an active part of crime statistics infrastructure to date and their use in measuring crime is unfamiliar terrain, and not because they are more important in any sense than the more traditional methodologies. It is also true that these new data resources may prove to be effective measures of some older, more traditional offense types as well.

Generally, new data resources must be harnessed because there are many crime offense types for which neither police-report data nor survey data are apt or workable as a source of offense counts and characteristics. One obvious example are crimes against governments and businesses that are not specifically spatial in a way that is linked to a local police jurisdiction. If a particular convenience store is robbed, the local police will likely be contacted; if a corporation's intellectual property is stolen by hackers attacking from overseas, the local police will not be involved. Indeed, the corporation may not notify any policing agency to avoid reputational damage done by public acknowledgement of its victimization. There are many other reasons that police report and victimization survey data are limited in what they can provide. For example, local law enforcement—traditional police—may either not have true jurisdiction relative to other regulatory agencies (e.g., environmental offenses) or simply may not be the first natural point of contact (e.g., fraudulent credit card transactions). In others, the crime manifests later in the sequence of events in the criminal justice system, such as the filing of charges or resolution through settlement or plea arrangement (e.g., corruption, bribery).

The panel's next conclusion returns to the base from which we work in suggesting the contours of national crime statistics going forward—the classification of crime developed in our Report 1:

> **Conclusion 2.4:** The classification of crime for statistical purposes recommended in Report 1 should be used as a blueprint for the nation's crime statistics. It should be used to guide revised measures of crimes currently included in existing data as well as pilot/initial data collection of as-yet-unmeasured crime types—even though it is impossible today for data collection to include every crime in the classification. While it is imperative that the classification undergo regular update and revision, initial data collection should begin with this as a base, rather than continue to refine the classification in the absence of data.

COORDINATION AND GOVERNANCE

The task of creating or modernizing any one of the subsystems we envision as part of modern U.S. crime statistics is difficult; implementing them all is harder still. And a guiding precept in constructing the classification of crime is that it should evolve, being revised over time to keep up with stakeholder needs and the changing nature of crime itself. In the panel's view, the essential point is to get started, with some measures and refinements, and not let the perfect be the enemy of the good. Real change cannot happen productively until first steps are taken and prototype measures are produced and scrutinized.

We gave extensive consideration to the factors that led to much slower adoption of NIBRS-format reporting than was anticipated—factors that affect the successful implementation of changes to the collection and content of crime statistics in general. In our assessment, the single greatest barrier to change in the nation's crime statistics system is there is no strong sense of ownership, or ultimate responsibility for, those national crime statistics. Both the FBI and BJS have strong, clear stakes in crime statistics, but the culture that evolved after 1930 was a relatively passive, hands-off mode of leadership at the national level. In important ways, this style is entirely predictable and understandable—stemming as it does from the reservation of most policing and criminal justice functions to the states under the U.S. federated system (and not the federal government), and also flowing out of abundant respect for the voluntary nature of contribution of data to national statistics, whether that is local law enforcement agencies filing UCR/NIBRS data or respondents completing an NCVS interview. But legal inability to impose on crime data providers and deference to voluntary contribution has led to a reluctance to motivate or incentivize crime data provision in any way, whether through "carrots" (e.g., direct technical-support funding) or "sticks" (e.g., linking data-reporting completeness to grant fund eligibility). Arguments can be made on all sides, but the end result is a crime statistics system that lacks essential leadership.

Specifically, we note that there is neither a strong coordinating body nor a defined governance structure for the whole U.S. crime statistics enterprise, dedicated to the generation of data on all types of crime. There is no reason why these two functions, coordination and governance, need to be conducted either by the same agency or by two different ones; we distinguish the two roles because there are subtle differences in what we see as the ideal roles in both areas. A strong national-level coordinating agency for crime statistics would be more actively and directly involved in the routine daily tasks of data collection and processing; it would have to cultivate and maintain an honest-broker role in interacting with data providers as well as providers of computer hardware/software solutions. The chief entity in a governance structure would take the broader view; its duties would include setting policy for the system as a whole, establishing the process for updating and maintaining the underlying classification of crime, providing a voice for and attention to feedback from the range of crime data stakeholders, and facilitating ongoing methodological research and development.

Two more conclusions by the panel note the absence of both these functions and make the direct appeal for them to be created:

> **Conclusion 3.1:** A stronger federal coordination role is needed in the production of the nation's crime statistics: providing resources for information systems development, working with software providers to implement standards, and shifting some burden of data standardization from respondents to the state and federal levels. The goal of this stronger role is to make crime data collection a product of routine operations.

> **Conclusion 3.2:** Having an effective governance structure for the complete U.S. crime statistics enterprise is critical. There is currently no entity responsible for reporting on the full range of crimes in the proposed classification (most notably for top-level categories 6–11).

The important issue of who should play these coordination and governance roles remains to be answered. Rather than make any recommendations about where the coordination and governance roles should be housed, the panel chose to identify requirements that any organization should fulfill in order to successfully carry out these two roles. What is critical is that national crime statistics be produced, coordinated, and governed pursuant to the expected sensibilities of a statistical agency, one for which data collection and dissemination—and attention to issues of data quality—are essential to the agency's mission. Specifically, this means adherence to the expected *Principles and Practices of a Federal Statistical Agency* that are regularly updated by the Committee on National Statistics of the National Academies of Sciences, Engineering, and Medicine (2017c).

Neither of the major agencies that currently collect crime data, BJS and the FBI, have all the necessary requirements, although both have some relevant attributes. BJS's authorizing language currently contains a very detailed list of specific duties—several of which are directly pertinent or responsive to the ideal coordination and governance roles. However, BJS has never been given the authority or been resourced in a way in which it could begin to fulfill all of them. Conversely, the legal authorization for the FBI's UCR Program is considerably more vague but has long endured and been supplemented over time. Addressing the new crime categories that have long been neglected requires paying attention to domains that fall outside the traditional highest priorities of either agency. Whatever agency (or agencies) serves as central coordinator for U.S. national crime statistics can only be effective if it can either work effectively with—or compel information resources from—myriad sources. This includes executive agencies spanning multiple problem domains and cabinet departments, independent agencies such as the Federal Trade Commission and the Environmental Protection Agency, quasi-independent entities such as the Postal Inspection Service, and federal and state law enforcement entities (including Offices of Inspector General). It is not clear if either BJS or the FBI is currently positioned to be successful in expanding the collection of crime statistics.

We are not aware of relevant research literature or other scientific footing that can serve as a basis for concluding that any particular organizational-chart approach is superior to another. Instead, we defer to the agency that may be best suited to manage the complex set of federal agency actors (and state and private authorities) who would ultimately be called on to contribute to crime data:

> ***Recommendation 3.1:*** **The U.S. Office of Management and Budget (OMB) should explore the range of coordination and governance processes for the complete U.S. crime statistics enterprise—including the "new" crime categories—and then establish such a structure. The structure must ensure that all of the component functions of generating crime statistics are conducted in concordance with the sensibilities, principles, and practices of a statistical agency. It should provide for user and stakeholder involvement in the process of refining**

and updating the underlying classification of crime. The new governance process also needs to take responsibility for the dissemination of data products, including the production of a new form of *Crime in the United States* that includes the "new" crime categories.

GOING FORWARD

Underscoring the statement made early in this summary, that crime statistics should not be limited only to broad counts of offenses, this report also argues that modern crime statistics require consideration of completely different ways to measure crime (in addition to improving the ways that it is counted or its level is assessed). We conclude that alternative metrics such as the harm imposed by or costs associated with crime merit research and, eventually, measurement and reporting:

> **Conclusion 2.2:** There is a problematic lack of solid information about the relative magnitude of the costs, harms, or importance across crime offense types or specific incidents. Further research could be sponsored on estimating the magnitude of societal harm associated with the major categories in the proposed classification—ideally, in a way that facilitates comparison over time and offense types.

Appendix D of this report includes longer-form discussions of some additional important topics, mentioned in brief in the main text but fuller consideration of which might otherwise disrupt the narrative flow. These include:

- Examining the commonly cited barriers that impeded adoption of NIBRS for 3 decades, including discussion of what has changed now to make success more likely;
- Arguing that improved data quality should be the major goal of voluntary contribution of data to national crime statistics;
- Observing that—rather than warn against overly simplistic comparisons of crime rates across jurisdictions and time—the dissemination platform for modern crime statistics should also provide easy access to the kind of contextual and community data that enables effective comparison; and
- Describing the challenges of getting a better reading on crime involving businesses and organizations as actors.

The path ahead for U.S. national crime statistics is neither easy, inexpensive, nor short in terms of time to completion. To modernize crime statistics will require a large investment in collecting data on things that are currently all but ignored in current crime measurement efforts. An adequately resourced

NCVS and a full-participation (and fully representative) NIBRS are important steps. Real benefit will arise from utilizing these tools and finally starting to shed light on long-neglected offenses, consistent with a rigorous classification and drawing from third-party administrative-type data resources that have not typically been thought of as part of the nation's crime data infrastructure. That data infrastructure should enable longer-run research and development on both new auxiliary measures of crime (such as the harms done by offenses) and on innovative methodologies for measuring offenses that elude detection in the administrative-type data resources. Although modernizing crime statistics will be challenging, it is critical to encourage these efforts. Crime and societal responses to all forms of crime result in substantial costs to the nation. Improved measures of crime are imperative for establishing policy priorities, evaluating policy effectiveness, and providing accountability data for agencies charged with responding to crime. There is great benefit to be realized from better statistics on the nation's crime problems; it is time to develop a more inclusive picture of crime in the United States.

– 1 –

Introduction

THE CURRENT, NATIONAL CRIME DATA INFRASTRUCTURE rests on two principal pillars, the Uniform Crime Reporting (UCR) Program administered by the Federal Bureau of Investigation (FBI) and the National Crime Victimization Survey (NCVS) sponsored by the Bureau of Justice Statistics (BJS). Beginning operations in 1930 following the guidance and structures outlined by a working group of the International Association of Chiefs of Police (1929), the UCR Program has morphed into a set of data collections that operate on a principle of voluntary contribution of information from local law enforcement agencies, nearly all channeled through a network of state-level coordinating agencies, for aggregation by the FBI. Within the set of UCR collections, two are central: the Summary Reporting System (termed SRS or UCR Summary in this report) that directly carries on the data series created in 1930, compiling counts or tallies of offenses reported to or otherwise known by law enforcement agencies, and the newer National Incident-Based Reporting System (NIBRS) that began operations in the late 1980s and that captures incident-specific detail on a larger set of offenses than the UCR Summary.[1] The second pillar of current national crime statistics is the NCVS, a household survey that—by eliciting both count and detailed

[1]As in the panel's Report 1, we maintain an important distinction between two basic units of measurement related to crime, between *offenses* and *incidents*. Offenses are the specific actions or behaviors—generally deemed socially unacceptable or subject to some sanction or punishment—that are partitioned into unique categories in one part of the crime classification recommended in Report 1. Incidents are events that can comprise one or more offenses that happen during that particular instance of "crime." A premise of our arguments is that ideal data collection should identify all of the offenses in an incident rather than collapse detail within an incident to a single offense (that is deemed to be most important or severe) as is done by the SRS.

information on personal victimization experiences—is positioned to generate estimates of crime offenses whether they are reported to law enforcement or not. The NCVS began operations in the early 1970s as the National Crime Survey and took its current name in 1992 after a thorough redesign.

The UCR Summary still adheres heavily to definitions and concepts laid down in 1929, with relatively little change—and, to important degrees (and for important purposes, including basic comparison across the sources), the conventions of NIBRS and NCVS were patterned after the UCR Summary at their formation. Thus, the current data collection mechanisms are relatively ill-suited—by construction—to cover new or emerging types of crime and so their limitations cannot be addressed simply through more complete collection of the same information. Some 30 years after NIBRS data collection began, and for a variety of reasons, NIBRS-format data are reported to the FBI from only roughly one-third of jurisdictions; likewise, the NCVS is enduring the same challenges as other federal surveys, trying to achieve reasonable response rates within tight cost constraints. The point is that even if these barriers to respondent participation were suddenly lifted—if the nation's current major crime data sources were to continue exactly as-is, just with full cooperation—the major gaps in content knowledge would remain poorly covered, simply because the topics concern offenses and require contextual information that the NIBRS and NCVS are not currently equipped to address in their current configurations.

The only way to start to answer these and other important questions about crime in the United States in a comprehensive way is to fundamentally modernize and redesign the underlying crime data infrastructure. It is amidst this mix of factors—the hardening of concepts and concepts of crime measurement over decades of practice, and the recognition of major, substantive gaps in the nation's quantitative understanding of crime—that the Panel on Modernizing the Nation's Crime Statistics has carried out its work.

1.1 THE PANEL AND ITS CHARGE

This panel's origin is described more completely in our first report (National Academies of Sciences, Engineering, and Medicine, 2016:Ch. 1; hereafter referred to as Report 1), but to recap in brief: This study is a successor to and extension of a previous, comprehensive Panel to Review the Programs of the Bureau of Justice Statistics conducted between 2007 and 2009. The BJS Review panel produced a first report (National Research Council, 2008) on options for conducting the NCVS—the bureau's most costly but one of its most important data collections—before focusing on the overall BJS portfolio in its final report (National Research Council, 2009). That review naturally touched on the UCR Program as a logical counterpart to the NCVS and given the BJS's legislatively assigned task of providing comparable data series of crime. However, that study was focused squarely on the BJS; more direct assessment of the UCR and its component programs, and the manner of "crime statistics" beyond the purview of the BJS (or the FBI), was explicitly outside its study scope.

This study was launched at the National Academies of Sciences, Engineering, and Medicine when the U.S. Office of Management and Budget (OMB) encouraged the BJS and the FBI to jointly sponsor this work—with the direct mandate to take a full view of the entire enterprise of data collection on "crime" in the United States and assess its relevance to modern information needs. The full statement of task of the current Panel on Modernizing the Nation's Crime Statistics is presented in Appendix A; developed in collaboration with all parties (BJS, FBI, OMB, and the Academies), it is a detailed yet broad mandate, to cover three main topic planks in two operational phases.

Our Report 1 (National Academies of Sciences, Engineering, and Medicine, 2016) was designed to cover the first, *substantive* plank in the panel's broader charge. The core of Report 1 is the presentation of a proposed, new classification for criminal offenses; said classification is a detailed revision of an International Classification of Crime for Statistical Purposes, maintained by the United Nations Office on Drugs and Crime and to which the panel remains indebted for providing a solid starting point. Our study was deliberately constructed in this fashion—starting with a classification system and only then moving into the full range of *methodological* and *implementation* issues, in this report—because it has been many decades since core concepts of U.S. crime statistics have been reviewed in full and from first principles. Traditionally, classification in U.S. crime statistics has been treated solely as a post hoc labeling and editing procedure; this study's basic premise is that there is considerable advantage in achieving clarity in concept first and building new indicators and data systems using the classification as a blueprint, rather than forcing existing statistics into a new post-processing mold. Another main objective of this final report, referenced in the full statement of charge and combining methodological and implementation elements, is to consider the appropriate governance structure—the most productive division of labor between the BJS and FBI, other data providers, (volunteer) law-enforcement-agency contributors, and other entities.

The centerpiece of the panel's data-gathering work, particularly for the first phase, was the convening of two workshop-style sessions in June and July 2014—supplemented by presentations and discussions at the panel's subsequent meetings in 2015 and 2016. While the workshop-style sessions were primarily intended to elicit user and stakeholder views on the need for crime statistics, they naturally and inevitably raised central implementation concerns as well, and so provided a trove of information underlying this report.

1.2 REDEFINING CRIME

Collectively, the panel's two reports represent a complete reconceptualization and redefinition of crime, in the context of U.S. crime statistics. At the outset of this report on methodological and implementation issues, it is useful to restate some basic precepts from Report 1 to explain our approach—in particular, ideas that we expressed in detail in Section 1.2 of Report 1's examination of "What Is Crime?"

At the outset in 1929–1930, the UCR Program established definitions based on the basic notion of "crime" as actions that are unlawful (counter to the language of criminal or penal codes) and so subject to punishment or sanction. This literal notion of crime motivated the initial restriction of UCR content to a small number of offenses, precisely because these major violent and property crimes were thought to be defined relatively consistently across the criminal codes of the various U.S. states (and because they were believed to be relatively well reported to local law enforcement agencies). However, consistent with our charge to adopt a broad perspective, we favor a behavioral concept of crime, defining it as a class of socially unacceptable behavior that directly harms or threatens harm to others. In Report 1, we argue in greater detail that the behavioral definition is less constraining and more dynamic in nature than the letter of legal text, befitting a phenomenon in crime that is constantly evolving based on shifts in technology and social norms. We also note in Report 1 that it is impossible to completely sever ties between the text of federal and state criminal code and our behavior-based definitions, and that the code remains an essential anchor. This is the reason why we exclude some behaviors that might be rude or unpleasant but that do not directly cause or threaten harm. It is also the reason why we insert the word "unlawful" in many of our specific offense distributions, because activities such as gambling or use/possession of some controlled substances are not criminal in all jurisdictions.

This broader perspective on crime differs from decades of crime statistics practice, and so it may be jarring to some. In particular, it necessarily blurs the line between things that are literally dubbed "crime" or "criminal" because they are adjudicated in criminal courts and those that are tortious or civil in nature (and so are adjudicated in other courts or proceedings). But this is the fundamental point of this panel's work; we will argue in this report that the nation's picture of crime is limited unless it considers *all* types of crime. Another critical point that should be made in this regard is that the task of classifying crime in Report 1 was to divide the entire space of behaviors that could be considered crime into unique offense categories; it was *not* to rank order them by importance, severity, or any other such scale. Homicide, robbery, harassment, environmental dumping, and tax evasion are all individual categories in the crime classification—and so we view each of them as a crime offense—but it would be a serious error to interpret our work as suggesting that they are equally important or that any of them is wholly unimportant relative to the others. Again, the interest is in a complete picture of crime in the United States, and that picture is incomplete without all perspectives.

1.3 LIMITATIONS OF THIS STUDY

In order to conduct its work, the panel had to make assumptions, recognize limitations, and make decisions on approach up-front. In order to calibrate expectations, we note several of these in this section as preface for the arguments that follow.

To begin, the role of this report is not and cannot be to serve as a precise, intricately detailed implementation plan. Rather, its role is to be a general blueprint—to speak to the shape, the contours, and the basic principles of a modern crime data infrastructure, and suggest general paths for implementation. This is the case for a variety of reasons, not least of which is that panels of the National Academies of Sciences, Engineering, and Medicine are generally precluded from describing or advocating specific federal budgetary priorities or recommending specific bureaucratic adjustments unless explicitly tasked to do so. Another important reason is a theme that we return to throughout this report: that there is formidable inertia in the enterprise of national crime statistics, and that the imperative is to make a start and refine based on the results rather than become mired in initial details. We cannot presume that our collective perspective, or even the wide range of user and stakeholder perspectives we sought in our meetings, is sufficiently thorough as to prescribe what offense measures should be improved first or what alternative data sources are more or less reliable than others. Arriving at an ideal and improved set of measures will require extensive iteration and refinement, based on prototype and preliminary measurement attempts and reflecting an even broader sweep of stakeholder input. We understand that some will look at our guidance in this report and wish for more specificity, but we conclude it to be best and most appropriate to maintain a fairly high-level view in suggesting revision of the crime statistics enterprise.

Another reason for not being too prescriptive or detailed on any specific point of implementation or methodology is, itself, a general constraint of this study. This constraint owes directly to the sheer extent of the panel's charge and the vastness and continual evolution of crime itself: We are obliged to emphasize breadth over depth on any specific area of crime or aspect of methodology. The panel's workshop-style sessions, the presentations at meetings, our deliberations, and our readings and research covered an extensive amount of terrain—and yet barely scratched the surface of what *could* be discussed. Some topics that we discuss at a high level—for instance, illegal drugs (and data on them), the relationship between firearms and crime, and the measure of elder abuse and mistreatment—have been the focus of panel studies in the past, and any number of others are ripe for in-depth review on their own.

Relatedly, the panel had to deliberately keep its focus on *crime* statistics—the material for crafting measures of levels and trends in the occurrence of specific offenses—as contrasted with *justice* statistics writ larger. The primary front on which we faced this challenge was addressed in Report 1, which defined what is meant by crime while minding the traditional distinction between what is typically processed in civil justice (or purely regulatory) proceedings versus criminal justice proceedings. There, we established definitions based on behaviors rather than statutory code language—arguably blurring the functional civil/criminal line in some instances but all in service of identifying behavior that should properly be labeled "crime." But there are further ramifications of a purely crime-statistics focus in this implementation- and methodological-centric report:

- The UCR Program has evolved over the years to become a set of data collections, as described in Box 1.1. Some of these are specialized crime/offense collections and so are germane to crime statistics; these include the UCR data collections on hate crimes and arson, as well as the Supplementary Homicide Reports (SHR). But the UCR's collection of data on arrests and clearances of arrests/cases—intended more as a measure of law enforcement activity and effectiveness, and of characteristics of the arrestee population, than of crime occurrence—and its collection of data on staffing and resources available to local law enforcement agencies are issues that we concluded out of our scope. This is not to exclude the possibility that arrests might be the most suitable or salient measure for certain specific offense types, because arrest might be the earliest measurable outcome in justice-system processing for those offenses. This might include some white-collar offenses such as embezzlement; we discuss such matters further in Chapter 2.
- Similarly, a major issue that has vaulted to the forefront in public debate during the lifetime of this study is data on the use of force by law enforcement, particularly resulting in injury or death of arrestees or suspects. Both the BJS and the FBI offered ideas for the generation of systematic use-of-force data as a standalone collection; at this writing, the FBI's proposed collection from local law enforcement partners appears to be going forward with the support of the Justice Department. But, having discussed the concept of justifiable homicide (and its role relative to our proposed crime classification) in Report 1, we do not believe it necessary to comment further on the disposition of either a standalone database on use of force by or involving law enforcement, or on the UCR Program's existing Law Enforcement Officers Killed or Assaulted (LEOKA) collection.

Box 1.1 Component Forms and Data Collections in the Uniform Crime Reporting Program

- *Summary Reporting System (SRS), Offenses Known to Law Enforcement:* The core of the Summary Reporting System (SRS) is Return A—Monthly Return of Offenses Known to Police and the separate Supplement to Return A. The separate offense tallies (and estimated total value of property involved in offenses) required by Supplement to Return A are described in Box C.2.
- *Supplementary Homicide Reports (SHR):* A form of detailed incident-specific reporting in current use, even among SRS-reporting jurisdictions, the SHR requests additional information about every homicide incident tallied in Return A. Particularly critical variables collected by the SHR are the victim-offender relationship and additional detail on the method/weaponry used in the murder. A limited but SHR-"branded" special collection was deployed in January 1935 in large cities, and primarily informed a more carefully delineated definition of justifiable (or "excusable") homicide. The SHR in its ongoing, incident-based form was first fielded in 1960; the 1961 issue of *Crime in the United States* was able to summarize findings from 3,008 SHR filings (Federal Bureau of Investigation, 1961:12).
- *Special Offense Counts:* Four separate data collection forms apply to offenses added to UCR Part I coverage in (relatively) recent years or to a specific class of offenses:
  - Monthly Return of Arson Offenses Known to Law Enforcement
  - Cargo Theft Incident Report
  - Monthly Return of Human Trafficking Offenses
  - Hate Crime Incident Report
- *National Incident-Based Reporting System (NIBRS):* The eventual successor to the SRS, NIBRS is separately maintained and its specially formatted data separately gathered from participants/state coordinators.
- *Age, Sex, and Race (ASR) of Persons Arrested:* Counts by basic demographic group of "persons dealt with by the police" through arrest were always intended to be added to the UCR Program, but were deferred for several years in the interest of cultivating coverage of the basic "offenses known to the police" tallies. Instead, admittedly weak proxy information derived from the fingerprint index cards filed with the FBI became a staple of UCR Bulletins for almost two decades. After "several years" of "experimenting with different types of forms," ASR forms were ultimately deployed to city police departments in 1952 (still excluding rural areas), and the first results from the ASR collection were published in the second UCR Bulletin for 1952 (Federal Bureau of Investigation, 1953:110–111). Over time, the ASR collection was split into its current form—separate forms (and counts) for adult arrestees (18 years of age or older) and juvenile arrestees.
- *Administrative, Law Enforcement Employee Counts:* To provide some context on the scope of local law enforcement operations, UCR Bulletins first published an estimate of law enforcement employees per 1,000 population (for very large cities 250,000 population or greater versus smaller cities) in 1933 (Bureau of Investigation, 1933:9). This was expanded to a set of city-specific rates for large cities in 1934 and continued over the years, evolving into a more expansive tabulation covering smaller cities by 1945. Today, an annual snapshot of law personnel under the UCR umbrella continues, using a form entitled "Number of Full-Time Law Enforcement Employees as of October 31."
- *Law Enforcement Officers Killed or Assaulted (LEOKA):* Data on "police officers killed by criminals" were first collected semi-systematically under the UCR program through "special reports forwarded to the [FBI] covering the calendar year 1939 and in examining similar data for the 2 preceding years" (Federal Bureau of Investigation, 1940b:85). Rates of assaults on police officers, by city size and geographic region, were added to *Crime in the United States* for its 1960 issue.

- Our predecessor National Research Council (2009:Finding 3.1) panel reviewing BJS programs critiqued current U.S. justice statistics for taking an overly cross-sectional approach—collecting detailed data "on an institution-by-institution basis (police, courts, corrections)" but in a way that "makes it difficult or impossible to answer questions about the flow of individuals from arrest through eventual exit from the system." However, that panel also recognized that this approach was necessarily the case, noting a critical need for greater capacity for linkages within and between crime and justice datasets. Such linkages include connections through all the steps in processing in the criminal justice system—from police report through arrest through charges through adjudication through corrections processing—as well as linkages to external social and economic data, as would be ideal for studies of recidivism. The current panel's recommended classification of crime is certainly intended to foster agreement on definitions and units of measurement, which is one part of the general problem—but beyond that, we do not believe it our role to go further. We agree that it would be ideal to develop a system of identifiers that would make it easier to trace the progress of crime offenses and incidents (and suspects/offenders and cases in general) through each stage in the criminal justice system, but the construction and implementation of such is beyond our scope.
- Related to this latter point, the FBI's Criminal Justice Information Systems (CJIS) division is responsible for maintenance of national criminal history record databases as well as the UCR; criminal history records improvement is an area in which BJS and the FBI have historically cooperated, leveraging BJS's authority to issue grants that the FBI lacks. There is obviously overlap between definition and measurement of "crime" and the promulgation of "criminal history" records. But delving into the content of criminal history record databases and fully dealing with the issues related to transmitting, disclosing, and linking person-level records in multiple datasets are all major tasks worthy of detailed research and debate in their own right, and not subjects that this panel is constituted to address.

It should be noted that both the FBI and BJS continued to develop their data collections during 2017, yielding changes that may not be documented in this report because they occurred after the panel's active data-gathering phase.

Finally, in terms of style and execution, we follow the practice of Report 1 in deliberately being very sparing in our designation of formal findings or recommendations. We chose that approach in Report 1 to spotlight a single recommendation—the panel's proposed classification of crime for statistical purposes. This report offers considerable commentary related to past and future crime reporting processes, but the panel opted to focus on a small number of broad, main messages that we designate as formal conclusions and a recommendation, rather than dilute this guidance with a longer checklist of highly specific recommendations.

1.4 STRUCTURE OF THIS REPORT

Chapter 2 begins where we left off in Report 1, comparing the coverage of the classification of crime we recommended in our Report 1 to the current national crime statistics, and to NIBRS in particular. Based on that coverage match, we outline the basic structure of a new system of multiple data collection efforts for crime statistics, using our proposed classification as a base. Chapter 3 rounds out our formal findings and suggests directions for coordination and governance within a modernized crime statistics infrastructure.

As previously noted, the full text of the panel's detailed charge is reprinted in Appendix A. As important reference for a focus on implementation and methodology issues, this volume also includes some extensive appendixes. Appendix B highlights important themes and lessons in the evolution of current U.S. national crime statistics, a review that provides important context about retooling the crime data infrastructure. Appendix C is a longer exposition of a rough, substantive match between our proposed classification of crime and its associated set of attributes/contextual variables with the nation's existing crime data resources, particularly NIBRS and NCVS. Appendix D offers longer-form commentary on some important, remaining issues that crosscut issues of crime statistics coverage and governance, important issues but ones for which full discussion in the body text might be unduly disruptive to the flow of arguments. To inform meaningful discussion of the role of the states in crime data collection, Appendix E presents state-level profiles summarizing legal requirements for crime reporting defined in state law or regulation and the status of adopting incident-based crime reporting. A final, shorter Appendix F sounds a cautionary note from international experience, using modernization experiences in the United Kingdom to underscore that some of the changes we suggest are not simple panaceas.

– 2 –

"Traditional" and "New" Crime: Structuring a Modern Crime Statistics Enterprise

In this chapter, we resume where we left off at the end of Report 1 (National Academies of Sciences, Engineering, and Medicine, 2016), with a detailed recommended classification of crime (offenses) for statistical purposes (Appendix D of that volume). After considering the extent to which the proposed classification is in concordance with current nationally compiled crime statistics, we outline some of the fundamental challenges to assembling indicators of crime for the United States as a whole but in a way that meets user and stakeholder needs for disaggregation and comparison. We then build from the recommendations in Report 1 to suggest general contours of an improved national crime statistics program.

2.1 MATCHING NEW CLASSIFICATION TO CURRENT PRACTICE

The table and commentary in Appendix C illustrates our basic assessment of the match between categories in our proposed crime classification with the Uniform Crime Reporting (UCR) Program's Summary Reporting System (SRS) and, particularly, the National Incident-Based Reporting System (NIBRS) that is now poised to replace the SRS as the primary source of information on crimes reported or otherwise known to the police. Through this matching, we offer a rough assessment of how consistent our proposed definitions or concepts for a particular offense are with the UCR's concepts (roughly, whether they are generally closely consistent, partially consistent, or inconsistent/uncovered in the current data), and so whether the data (again, principally NIBRS) are capable of generating estimated counts or measures of specific offenses that may be disaggregated by geographic or demographic characteristics.

For completeness's sake, the "full match" in Appendix C also includes a column indicating the extent of consistency and coverage of the offense in the National Crime Victimization Survey (NCVS). We return to this point in other parts of this chapter and this report, but the basic features of the match to the NCVS warrants a comment up front. An unavoidable impression that one might draw from the table in the appendix is that the NCVS and our classification make for an exceptionally poor fit—but that is a very misleading impression. Most crime is a relatively rare event in the statistical sense, so the vast majority of the respondents to a household survey like the NCVS will not have victimization experiences to report for any particular offense. The sample size of the survey would have to be impractically large to achieve fine-grained detailed estimates, and so tabulations from the NCVS—by necessity—tend to stick to offenses that occur most frequently and to groupings of offenses (e.g., violent crime relative to property crime). The comment that is in order up-front is simply that the seeming mismatch between our classification and the NCVS should not be interpreted as criticism of the value of the NCVS. Indeed, as we will discuss later in this chapter, survey-based measures of crime are an integral part of the system of data collection systems we envision.

The information from the full match in Appendix C is conveyed in a more condensed, approachable form in Table 2.1; it focuses on the rough correspondence between our classification's major subcategories and concepts in UCR/NIBRS.

The most striking result of the match can be gleaned by glancing down the columns of Table 2.1: The coverage of our classification's categories in NIBRS is, overall, very sparse. This is not surprising, given that the express intent in Report 1 was to cast a very wide net in defining "crime" offenses through our classification, the intent of a statistical classification to be a partitioning of the entire range of the phenomenon in question. As in Report 1, we do not intend for this result to raise concerns that assembling modern crime statistics is a futile task—it is not—but rather to be candid about gaps in knowledge and to be aspirational in seeking a fuller picture of crime.

However, the match is also markedly differential in nature. It is worth reminding, at this point, that there is no gradation of importance or seriousness (or anything else) implied by the numbering of categories in our classification. Nonetheless, the numbering scheme is such that top-level categories 1–5 correspond to what might be considered "traditional" crime: the well-known,

Table 2.1 Summary, Match Between Panel's Proposed Classification and National Incident-Based Reporting System (NIBRS)

Category in Panel's Classification	Coverage in UCR/NIBRS: Covered	Partially Covered, or Different Concept	Not Covered
1 Acts leading to death or intending to cause death	Murder and intentional homicide Nonintentional homicide	—	Assisting or instigating suicide Unlawful euthanasia Unlawful feticide Unlawful killing associated with armed conflict
2 Acts causing harm or intending to cause harm to the person	Coercion	Assault Threat Acts against liberty Trafficking in persons Dangerous acts	Slavery and exploitation Negligent acts Acts intended to induce fear or emotional distress Defamation Discrimination Acts that trespass against the person
3 Injurious acts of a sexual nature	Rape	Sexual assault Sexual exploitation of adults	Sexual violations of a nonphysical nature Sexual exploitation of children
4 Acts of violence or threatened violence against a person that involve property	Robbery	—	Terroristic or disruptive threats to buildings or critical infrastructure
5 Acts against property only	Burglary Theft	Property damage Acts against computer systems (unlawful access to a computer system)	Acts against computer systems (all other) Intellectual property offenses

Table 2.1 (continued)

6 Acts involving controlled substances	—	Unlawful acts involving drug equipment or paraphernalia	Unlawful possession or use of controlled drugs for personal consumption Unlawful cultivation or production of controlled drugs Unlawful trafficking or distribution of controlled drugs
7 Acts involving fraud, deception, or corruption	Forgery or counterfeiting	Fraud (identity theft; false pretenses; mode-specific types such as wire fraud) Corruption Acts involving proceeds of crime	Fraud (generally, including fraud against businesses or establishments or government agencies)
8 Acts against public order and authority	—	Acts against public order sexual standards	Acts against public order behavioral standards Acts related to freedom/control of expression Acts contrary to public revenue or regulatory provisions Acts related to migration Acts against the justice system Acts related to democratic election Acts contrary to labor law Acts contrary to juvenile justice regulations or involving juveniles/minors

Table 2.1 (continued)

9 Acts against public safety and national security	—	—	Acts involving weapons, explosives, and other destructive materials Acts against national security Acts related to organized criminal groups Terrorism
10 Acts against the natural environment or against animals	—	Acts against animals	Acts that cause environmental pollution or degradation Acts involving the movement or dumping of waste Trade or possession of protected or prohibited species of fauna and flora Acts that result in the depletion or degradation of natural resources
11 Other criminal acts not elsewhere defined	—	—	Violations of military law Violations of tribal law Torture Piracy Genocide War crimes

NOTES: —, not applicable. Full version of match is presented in Appendix C.

violent interpersonal "street" crime and property offenses that dominate the nation's current crime statistics. Within these broad sectors, agreement between our classification's traditional crimes and NIBRS is reasonably good. That said, the middle "partial match" column for the traditional crimes significantly downplays the magnitude of differences in concept between our definitions and NIBRS's definitions—perhaps most importantly the radically different conceptualization of assault and threat that we postulated in our classification (dissolving the UCR concept of aggravated assault). By comparison, our categories 6–11 correspond roughly to "new" types of crime—new, in most cases, in the sense of being new in statistical tallies, having received little to no standardized national collection to date even though they have been serious problems for decades. In this range, there are very few entries in the pure "covered" column but a few hopeful signs in the "partially covered" column of Table 2.1. Category 6, acts involving controlled substances, is interesting middle ground—certainly, a group of offenses that the public might think are well documented by the nation's crime statistics. But, in current practice, the UCR's categorization of "drug/narcotic offenses" is sufficiently vague and the UCR measure (arrests only) is sufficiently detached from the full behavior of interest that we deem the match between NIBRS and our classification to be poor. Hence, it makes sense to include controlled substance offenses as "new" crime.

2.2 CONTOURS OF A MODERN NATIONAL CRIME STATISTICS INFRASTRUCTURE

Against this backdrop—lessons learned from decades of data collection under current structures and a conceptual framework for understanding what a complete picture of crime should look like—we now articulate our vision for a modern national crime statistics infrastructure. In doing so, we also render three formal conclusions, main messages that we hope to convey through this report. These three conclusions, and the balance of this chapter, can most succinctly be described as arguing that modernizing U.S. "crime statistics" demands attention to both parts of that phrase. Our Report 1 and recommended classification collectively argue that improved crime statistics require a broader concept of *crime*, and we return to this point with our third conclusion. But we begin with arguments that emphasize that there is a need for more *statistics* in modern crime statistics, too.

2.2.1 Measuring Crime, Not "Enumerating" It

Appendix B provides fuller historical background behind the following statements, but two damaging trends took hold during the decades in which the

UCR Summary system was effectively the sole source of nationally compiled crime statistics:

- *Undue sense of completeness and accuracy:* Having achieved high (and genuinely impressive) levels of voluntary contribution of data by local law enforcement agencies, the tenor of communications about the crime data changed. Near complete participation was tacitly characterized as universal participation—with hints about lapses in reporting by individual agencies (in whole or for some months in the year) being expressed not at all or at best in footnotes or auxiliary documents. Frank acknowledgments of limitations in the data gave way to broad, declarative summary statements and authoritative-appearing tables. Fundamentally, notions of uncertainty and error—for instance, the extent of imputation required due to agency nonresponse or the extent of measurement and conceptual error—dropped out of view. At the same time, the report *Crime in the United States* was largely taken to *define* "crime in the United States"—presented and commonly interpreted as an exhaustive enumeration of crime in the nation when in fact it never was and (limited to selected offense types and to offenses reported to law enforcement) it never could be.
- *Reduction to coarse, short-term changes in a few broad categories of crime rates:* When counts of offenses are all that are available, absent detail and context, attention naturally focuses where it can—on the year-to-year upticks or downticks of national level counts. While information on year-to-year changes has utility, it is not sufficient to more fully inform understanding of any changes in the nature of crime. It also diminishes the key potential value of nationally compiled crime statistics to the wide array of data users and stakeholders, for whom the overriding potential utility of the UCR data is comparison of a jurisdiction's crime conditions with those in other jurisdictions. The counts (absent context and understanding of possible errors within them) render such comparisons inapt and potentially misleading.

As also recounted in Appendix B, more recent years have seen new concepts or offense types emerging in national interest—arson, hate crime, human trafficking, cargo theft—forced into the count-intensive, bookkeeping-type UCR Summary mold by default.

Accordingly, a first principle for modern crime statistics is that coarse tallies are not enough to meet the needs of the full range of users and stakeholders. For any given offense type, the starting point of the discussion should always be the basic question: What does the nation need to know about crime, and how may it best be measured? A major part of the answer has to be acceptance that measurement of crime will necessarily involve uncertainty, which should be documented and scrutinized. This report is an argument for recasting the crime statistics problem—for "traditional" and "new" crimes alike—as one of measuring and estimating levels of and trends in crime and not simply (or only) counting it.

> **Conclusion 2.1:** The aim of modern crime statistics is the effective measurement and estimation of crime. Accurate counting of offenses and incidents is important, but the nation's crime statistics will remain inadequate unless they expand to include more than just simple tallies with no associated measure of uncertainty or capacity for disaggregation. Through the collection of associated attribute data, the crime statistics we suggest should—at minimum—enable the analysis of data in proper geographic, demographic, sociological, and economic context, and provide the raw material for important measures related to an offense (such as the harm it causes) in addition to its count.

There are numerous major issues packed into this call for a return to estimation in crime statistics; we discuss a few in the balance of this section, and more detailed treatments of others are developed in Appendix D.

Nondenumerable Crime

The last sentence of Conclusion 2.1 has application for some traditional crime offense types. As described in Appendix B, one of the earliest appeals for information beyond a total, aggregate incident count was for the monetary value of estimated property loss, particularly for the pure property crime of larceny-theft. That shortcoming led to the development of UCR Summary's Supplement to Return A. The exact manner by which that information is currently collected was heavily criticized in our panel's workshop-style sessions with crime data stakeholders. They found dubious the totaling of property values by offense type and crude (and somewhat dated) property type categories. By comparison, collection of detailed incident-level property loss data would be more useful—and arguably essential for understanding the dynamics of larceny—because of their capacity for aggregation and recombination.

Yet it is the "new" types of crime in categories 6–11 of our classification, especially, that lead us to state that last clause of Conclusion 2.1 as we do. This is because many of the new offenses defy the application of consistent counting rules—and, moreover, are effectively nondenumerable because raw counts may actually be misleading or meaningless absent availability of other information. Some examples may be useful to consider:

- *Deployment of ransomware*, malicious software that denies users access to their own data and computer systems until a ransom is paid electronically: Counts of ransomware deployments do relatively little to improve understanding of the severity of the problem. The deployment of a single ransomware script may be aimed at a small set of users or be propagated worldwide, touching vastly different numbers of downstream victims. Likewise, there are certainly material differences in the amount of damage caused by a single ransomware deployment that locks up an individual user's machine and one that hobbles use of the automated information systems of an entire hospital or medical group.

- *Environmental violations:* Technically, the same offense count (1) would apply to the improper disposal of a laptop computer battery as to a deliberate spill of large amounts of toxic materials into the water or air. For such offenses, noncount measures of interest may be crucial to consider, such as the amount of pollutant discharged into the environment, the percentage by which emissions exceed values permitted by regulation, or the size of the population that may be affected by the discharge. Even when the size of two offenses are similar, the consequences may not be. Spilling the same number of barrels of oil into a water supply may not be equivalent to spilling the same number of barrels of oil in a nature reserve or an industrial site.

Ultimately, the core component of crime statistics is the assessment of the level and characteristics of crime and trends in its occurrence, and so it stands to reason that some manner of count is essential. Again, Conclusion 2.1 underscores that which has already been said about crime statistics focused overly (or exclusively) on aggregate tallies, as in the UCR System—those coarse counts, incapable of disaggregation by factors of interest or isolated from other pertinent information that are more descriptive of a particular offense better than an incident tally, are simply insufficient in a modern information environment. Counts capable of disaggregation and detailed contextual analysis may be ideal for traditional offense types, but the new crimes in particular require supplemental information because counts alone are acutely insufficient for understanding and use.

Challenge of Timely Crime Data

Another aspect of measurement that presents challenges for national-level statistics of all kinds is timeliness. This can be difficult for the broad range of offense types in our classification, but especially for the crimes in categories 6–11. The important questions to consider for each offense in the classification is what does the nation need to know, and how best might it be measured, which invites the follow-up question of how the best measure is determined. The common adage that things can be done cheaper, faster, or better—but not all three simultaneously—rings true and is a constant tension. As we will argue in more detail later, our global preference is for better in the sense of accuracy and data quality, but timeliness can become a paramount issue for certain types of crime, such as homicide and violence. Timeliness regarding national-level crime statistics refers not only to the availability of crime data for analysis and policy decision-making, but also to conceptual decisions about when some crimes can be determined to have occurred.

The panel's Report 1 discussed some of the problems that are associated with the fact that national crime statistics typically are released by the FBI in its annual publication *Crime in the United States* approximately 10 months after the collection year ends. (For example, crime statistics for 2016 were released in October 2017.) Report 1 notes that in 2006, and again in 2015, national and local concerns about apparent homicide and crime increases were growing. Although police departments had information about their own local crime problems, the lack of timely UCR data for comparison purposes made it difficult for law enforcement to determine whether increases were unique to their own cities or part of a broader national pattern. In an effort to obtain data to evaluate these problems, police organizations could not turn to recent UCR data, so they surveyed their membership about local crime problems. News media, advocacy groups, and other organizations also began creating their own databases derived from available local police departments' website information. This nonsystematic data-gathering can be very problematic and can result in unproductive debates about the causes of the apparent increases and resource needs. Homicide increases are of great concern, and for this type of offense, more timely data release is critical.

Yet, for other types of offenses, the issue of timeliness involves different conceptual dimensions. For example, in conspiracy crimes, it may take time for the crime to manifest and be detected by law enforcement. In consumer fraud or credit card fraud, it may take time for even the victim of an offense to know that it has occurred. In these types of instances, there can be considerable time between the occurrence of an incident and the point at which it can be detected by survey or other methodologies. However, for these offense types, the issue of quick and timely data availability for comparison purposes is arguably less important than it is for crimes involving homicide and serious violence. The speed at which data can be made available is expected to vary across offense types, but when there is evidence that data are urgently needed for important decision-making, as in the case of homicide and serious violence, that information should be released more quickly.

Today, there is an ever-increasing desire for data that are available quickly, and technology itself has enabled construction of data resources with speed unthinkable in past years. Web-scraping methods that scour Internet pages, media accounts, or social media postings in search of information about a topic of interest have found some application in the analysis of crime. Media organizations and individual academic researchers are now maintaining their own repositories of data on homicides, shootings, or terrorism-related incidents. Another prominent, and relatively long-standing, example growing out of academic research is the Global Terrorism Database that mines media stories on terrorism-related events as its first-step input, and which has been characterized by the investigators as tapping into a "fire hose" of news media articles (Miller, 2014:6). The development of terrorism event databases and challenges encountered in the experience of the Global Terrorism Database are described by LaFree et al. (2015) and LaFree (forthcoming), but one lesson is that thoughtful conceptualization of the problem at hand and an infrastructure of data quality controls (e.g., verification) are critical to the reliability of these data. In other words, it is good to have an aspirational vision of data resources—on crime, or anything else—that are truly "real-time" in nature—but even real-time may involve reconciling occurrence time lags between when an offense occurred and when it became detectable. Moreover, they *should* involve some lag in order to satisfy basic data quality requirements.

It is instructive, at this point, to recall the familiar funnel model of operations in the criminal justice system (see Figure 2.1). The occurrence of crime is the wide, open end at left of the funnel and the later phases (investigation through adjudication and on through corrections) are the increasingly thinner segments progressing to the right. Given multiple possible measures of the same offense, the understandable general preference is for the one closest in time to the actual occurrence, and so the one that falls leftmost in the crime system funnel—the earliest point at which an occurrence is reported by a victim or observed by actors in the justice system. But this is not always possible, and may necessitate glimpses further to the right in the justice system funnel than may be desired, simply to make measurement possible. Suspicious fires typically require lengthy investigation before it can be determined whether the measurable behavior constitutes the offense of arson; in white collar offenses such as embezzlement and some types of fraud, it may be impossible to pinpoint exactly when the offending behavior occurred, and the earliest point at which the "crime" might be ruled to constitute an offense is either at the point of arrest or the filing of legal charges. In still others, such as corruption, bribery, and some intellectual property offenses, the best or only time for determination and measurement is later still, upon resolution through settlement, plea arrangement, or conviction.

Reconciling the time of occurrence of a crime can be very messy—all the more reason, as suggested by Conclusion 2.1, to take care to position crime statistics as estimates and not as authoritative point-in-time enumerations.

2.2.2 Putting Crime in Context, Alternative Metrics for Crime, and the Harms Imposed by Crime

A recurrent theme throughout our panel's work is that modern crime statistics would be more useful if they can provide more information and more

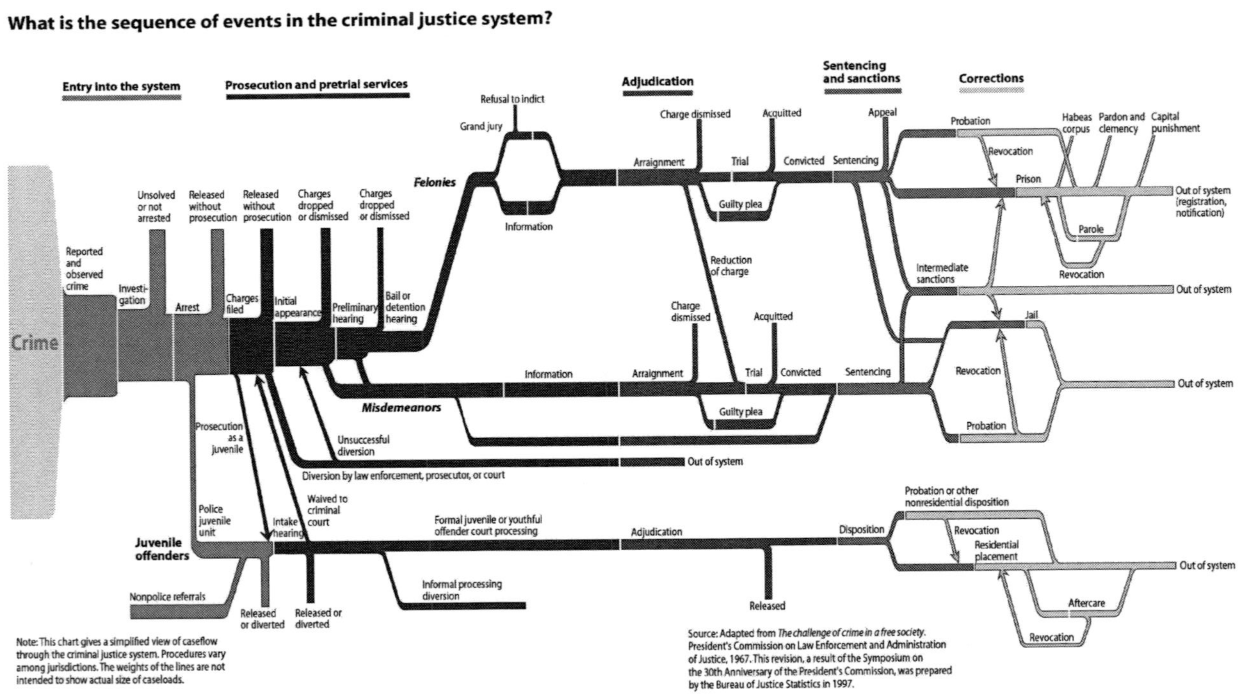

Figure 2.1 Sequence of events in the criminal justice system.

SOURCES: Adapted from President's Commission on Law Enforcement and Administration of Justice (1967:8–9) in 1997; posted at http://www.bjs.gov/content/justsys.cfm.

flexibility for detailed analysis than simple counts and summaries alone. In the next section, we turn to the specific data collection systems that we expect to populate national crime statistics going forward, Two important points should be made about information that it is supplementary to traditional measures of crime. The first is that users can only draw meaningful conclusions from comparisons of crime data across jurisdictions or across time if they have the information necessary to put crime in appropriate geographic, demographic, and other context. This is one reason for the attributes that we include as a part of our proposed classification, but the argument extends beyond that. Part of national crime statistics going forward should be ready provision of applicable *noncrime* data that can be used, side-by-side and in concert with levels of crime (however offenses might be measured), to draw effective conclusions. This issue, including the work of a Crime Indicators Working Group commissioned by the Bureau of Justice Statistics, is discussed in more depth in Section D.3.

The second essential point is that entirely different measurements—not just trying to quantify the occurrence of crime—may be most useful or salient to understanding crime. These alternative metrics may include estimating the harm done by or the cost of crime. In defining "crime" in Section 1.2 of Report 1, we noted that "there are other measures related to offense behaviors (such as estimates of damage or financial loss inflicted, or even the *perception* of victimizations or occurrences) that may have greater salience" than counts for some offense types. We broadened that concept in the last sentence of Conclusion 2.1 in this volume, arguing that new and detailed crime statistics should "provide the raw material for more useful measures related to an offense (such as the harm it causes) in addition to its count."

Harm is inextricably linked to crime. In Report 1, we took as a basic behavioral definition of crime that it is "a class of socially unacceptable behavior that harms or threatens harm to others," and this behavioral definition was the basis for all the ensuing specific definitions of offenses in our classification. To make that definitional task tractable, we had to treat harm as a binary, yes-or-no concept—as, for example, in drawing a sharp and observable difference between "assault" and "threat" being the actual infliction of bodily injury, or by resorting to "unlawful" as the criterion by which to indicate whether a toxic substance is sufficiently harmful as to be branded an "environmental offense." There are a few exceptions to this treatment of harm as a binary indicator for definitional purposes. For example, our revised assault category also includes a definition for serious bodily injury to facilitate a distinction between minor and serious assault, and our breakdown of category 6, controlled substance offenses, is premised in part on the quantity of drug involved (and so the amount of harm that could be wrought).

The point of this current section is to emphasize that harm is not binary, and that the measurement of harm associated with crime is an important consideration in its own right. Yet it is also true that measuring the harm associated with crime is both difficult and contentious—a major research effort in its own right, and one that might not be a cleanly resolvable problem. As this report was in final development, the U.S. Government Accountability Office (2017) issued a report summarizing its findings from a survey of 16 experts on estimating the cost associated with crime, summarizing several approaches and corroborating the lack of agreement on a best approach to the problem.

Although efforts to measure the costs of crime are known to be challenging, there are critical reasons for encouraging these efforts. Crime and societal responses to crime result in substantial costs to the nation. Adequate measures of the harms from crime and victimization are necessary for evaluations of some crime control policies (Greenfield and Paoli, 2013; Paoli and Greenfield, 2015). These types of measures are critical for establishing policy priorities across a wide range of offense types. They would also allow for evaluations of policy effectiveness for specific offenses, as well as better accountability from agencies charged with responding to crime. To achieve these important goals, research and development of estimates of the harms and costs of crime are essential.

In this report, we are very sparing in formally designating conclusions and recommendations. We offer many suggestions along the way but have preferred to keep the formal-language focus on a small number of main conclusions. Alternate metrics of crime should be an important part of a modern crime statistics system that is not limited solely to offense counts, but full exploration of the alternatives is simply not possible in this study with its already expansive mandate. This concept is sufficiently different from the other guidance in the report that we think it appropriate to designate it as a formal conclusion, as follows:

> **Conclusion 2.2:** There is a problematic lack of solid information about the relative magnitude of the costs, harms, or importance across crime offense types or specific incidents. Further research could be sponsored on estimating the magnitude of societal harm associated with the major categories in the proposed classification—ideally, in a way that facilitates comparison over time and offense types.

Offense counts alone are almost certainly not the correct measure of crime and, as we have already noted, the new types of crime in our classification in particular (but also some of the traditional offenses) will require further, dedicated thought regarding the most appropriate and accurate measure of level or trend—whether report to law enforcement, report to a survey interviewer, arrest, conviction/ruling/imposition of sanctions, or the like. It will necessarily take some trial and error to develop the right mix of such indicators. It will also take time to determine the kinds of alternate, auxiliary measures that may add much to national understanding of crime.

2.2.3 New Systems for New and Traditional Crimes

The second main message of this report builds directly from Conclusion 3.1 in Report 1, in which we concluded that:

> No single data collection can completely fulfill the needs of every user and stakeholder, providing data with sufficient detail, timeliness, and quality to address every interest of importance. Any structure devised to measure "crime in the United States" should necessarily be conceptualized as a system of data collection efforts, and informative details about the collection and quality of the distinct components in this data system should be included to help ensure proper interpretation and use of the data.

Specifically, it reinforces the system-of-systems concept for a modern crime measurement infrastructure, stipulating a continued role for the current core methodologies but vitally extending them as well. Thus, we reach the following conclusion:

> **Conclusion 2.3:** Improvement in the nation's crime statistics will require enhancements to and expansions of the current data collections, as well as new data collection systems for the historically neglected crime types highlighted by the proposed crime classification.

In practice, this means that the set of output crime statistics expected from a new crime data infrastructure should strive for uniformity in concept and underlying definition—so as to permit detailed analysis and comparison (in appropriate context) across jurisdictions and over time. But the statistics need not be completely uniform in mode of measurement for all offense types, depending on what is deemed to be the most effective means to measure the requisite information about specific offense categories. The dual statement to the argument that "no single data collection can completely fulfill all needs" is that no single data collection should fulfill (or be made to fulfill) those needs, either. Specifically, the three component data systems that we envision are:

- an incident-based recording system covering offenses known to law enforcement agencies;
- a survey data component, principally the National Crime Victimization Survey (NCVS) and its topic supplements, but also including other surveys as appropriate, including an emerging BJS survey intended to improve the measurement of the specific offenses of rape and sexual assault (see National Research Council, 2014); and
- a variety of primarily administrative-record-type data sources, primarily for coverage of new crime types that are outside the scope of either police-report or household survey methods.

In Report 1, we also argued (Conclusion 5.4) that:

> Full-scale adoption of incident-based crime reporting by all respondents or sources, that is sufficiently detailed to permit accurate classification and extensive disaggregation[1] and analysis, is essential to achieving the kind of flexibility in crime statistics afforded by a modern crime classification.

The phrase "all respondents or sources" is included precisely because multiple component systems are necessary within a broader, new crime statistics infrastructure. As we note below, this language can fairly and accurately be read as strong support for the National Crime Statistics Exchange (NCS-X) initiative and progress toward full-scale implementation of NIBRS, as well as the basic structure of NCVS interviews (eliciting incident-specific detail from its respondents). In both the survey and administrative-records setting, the precise application of the attributes or contextual tags specified as part of our classification will depend on the specific crime because not every attribute is necessarily meaningful or appropriate for every possible offense. But the spirit of the conclusion remains the same: The key point is to avoid the sole collection of monolithic counts, as in the UCR Summary system, and collect detailed incident-specific data to the extent practicable.

New System for Police-Report Data: NIBRS as First But Intermediate Step, and Switching from Reporting to Recording

The first major component of the modern crime statistics infrastructure we recommend is an incident-based recording system covering offenses reported to or otherwise known by law enforcement agencies. The basic nature of crime is such that police-report data will continue to be an essential resource. Indeed, it will be the principal source of statistical series on the traditional crimes identified in our classification, as well as a good percentage of the new crimes. This data system could also be characterized, as we did in Report 1, as being "attribute-driven," referring to the set of attributes or contextual tags that we posited as constituting an essential part of our recommended classification and a minimum level of detail desired for eventual disaggregation and data analysis.

We strongly support the BJS's National Crime Statistics Exchange (NCS-X) initiative, NCS-X's cultivation of a nationally representative sample of NIBRS-reporting jurisdictions, and the sunset of the UCR Summary Program in favor of a fully operational and full-compliance NIBRS. But there are two principal reasons why we did not simply designate NIBRS as the first piece of our envisioned modern data infrastructure.

[1]By "disaggregation," we refer to having more than aggregate summary counts for individual categories, but also making use of the attribute/contextual information included in our classification to permit use-driven breakdowns of offenses and incidents into more meaningful analytical subsets.

The first is that *even a fully realized NIBRS is not enough to provide necessary information on crime and to reconcile key conceptual mismatches.* In closing Report 1, we described NIBRS as an "intermediate step" in getting to the kind of data collection that we desire, intermediate because the nation's comprehensive picture of crime would be incomplete even if full NIBRS participation were achieved today. However, the conclusion that the picture is incomplete because of the sheer scope of covered offenses between our proposed classification and NIBRS would be mistaken. We intend for police-report data to be the principal source for those offenses for which those data and that methodology are the most apt—not the entire classification. Moreover, as described in Report 1, we deliberately limited the attributes (contextual information) that are part of the data collection to things that are relatively easy to determine objectively and quickly—and, pointedly, to keep attribute data collection at roughly the same level of effort required to assemble NIBRS's data elements. Hence, in reconceptualizing police-report crime data collection, we have tried to keep the request realistic—some expansion of covered offenses relative to NIBRS, to be sure, and some revision of attributes/contextual elements. But, going forward, the principal substantive limitation with NIBRS relative to our suggestions is the major difference in general concept and definition for some major offense types—the things that led us to conclude only a "partial match" between our classification and NIBRS coverage. In developing our classification, we reached the firm conclusion that the UCR/NIBRS concept of aggravated assault is fundamentally flawed and that better information would be generated by retooling the concepts of assault and threat; likewise, the basic differences in concept concerning other crime types, such as sexual assault, harassment and stalking, and fraud, among others, are such that a complete NIBRS only provides part of the information that the nation needs.

The second reason for not stopping with NIBRS is that—mindful of the voluntary nature of police-report data—the timing is right to fundamentally *change the focus of police-report data from reporting to recording, from preparation of specialized products to direct transfer of data as a seamless by-product of normal operations.* The goal of mounting a voluntary data contribution system that is effectively an invisible by-product of day-to-day local police agency operations, requiring minimal if any direct effort, was simply not attainable in the era of the UCR Summary System. Under the guise of providing simple summary counts, the process of completing and submitting the monthly Return A and Supplement to Return A required a very intricate bookkeeping system—effectively maintaining ledgers of offense tallies, arrests, clearances of arrests, property of stolen goods and so forth, not to mention the filing of supplemental returns (e.g., a −1 and a +1 to account for a serious assault converted to a homicide by the death of the victim). Meanwhile, the cognitive and processing burden of ensuring compliance with UCR definitions and concepts was placed squarely on the local law enforcement responders themselves, making completion of even the "simple" Return A by a smaller agency (without a robust automated information system) a potentially overwhelming task.[2]

We are mindful (from Appendix B's historical review) that, from the outset in 1930, planners of national crime statistics have frequently severely underestimated the extent, quality, and ease of conversion to requested data formats of local law enforcement agency records, and we do not make such an assumption casually now. Also, there is certainly variation across agencies in the expected rigor and format of officers' reports (and the extent to which officers are actively trained, or not, in report writing), including the possible role of those reports in downstream local and national crime statistics. Finally, sensitivity also exists because a "finished" officers' incident report may actually still be a "live" investigative record—all factors that may affect the quality of reports entered into a local agency's records. Yet at the same time, that first responder's report may be the first or best chance at ascertaining such factors as bias or "hate crime" motivation in an incident, or accurately capturing other attributes of a particular offense.

We will return to these critically important issues in Section D.1 (Appendix D), in which we delve into them in depth that would otherwise disrupt the flow of argument here. All that said, our interactions with a wide variety of crime data stakeholders and local law enforcement officials strongly suggest that the days of local police records being hard-copy paper reports stored in file cabinets are numbered, if not over. The pervasiveness of automated information systems in usage by law enforcement agencies, for support of all functions from officer dispatch through the investigative process, was something that was underestimated by NIBRS planners in the 1980s and 1990s. In 2017, as we understand from our discussions with agency practitioners, it has become the norm for all but a very few local law enforcement agencies. Technological advances enable the long-hoped-for vision of a data system premised on direct transfer of records to state and then to national repositories, shifting much of the burden of editing, standardization of concept, and ensuring data quality to the state and national levels.

Survey Data as Essential, Complementary Measures

Self-report, survey information on criminal victimization—including but not limited to an adequately resourced NCVS—should also remain a pillar of U.S. national crime statistics. Survey data on crime are necessarily limited in the degree to which the resulting data can be disaggregated along useful analytical

[2]Our workshop-style meetings turned up multiple NIBRS-reporting jurisdictions that had encountered an unexpected handicap: Their records management systems turned out to be incapable of directly computing UCR Summary Return A files. Instead, they had to rely on their state UCR coordinators to translate their incident-based files into Summary format.

lines. This is because crime is still, relatively speaking, a rare event in the population at large, so the survey's sample size would have to be unworkably large to support fine-grained estimates. But survey data should remain part of the crime statistics infrastructure because of their unique, conceptual strength—the ability to estimate the level of victimization that is *not* reported to law enforcement authorities. Survey-based measures can never, without prohibitive cost, achieve fine detail in both offense coverage and geographic/demographic group; their province will continue to be relatively high-frequency offenses and for relatively coarse population subgroups and at relatively low frequency (e.g., annually). But the value of an independent source of victimizations not reported to the police—as a check on estimates derived from either police-report data or administrative/external files—makes survey measures a vital part of the statistical infrastructure.

Accordingly, survey data are poised to serve most often as secondary measures of offense categories or groups of offense categories. However, it is entirely possible that, as individual offenses are reviewed for the most appropriate mode of measurement, the most salient and meaningful national statistics on some offense types may turn out to be NCVS-based measures (at the expense of geographic disaggregation). An iteration of the Identity Theft Supplement to the NCVS yielded estimates suggesting on the order of 16.6 million adult U.S. residents experiencing one or more incidents of identity theft in 2012—incurring estimated monetary losses on the order of $25 billion (Harrell and Langton, 2013). Tellingly, the Federal Trade Commission (FTC) partnered with BJS on the supplement (along with the Office for Victims of Crime)—in part, undoubtedly, with interest in calibrating the results with experiences reported to the FTC's fraud hotlines and in its own data collections. Other offenses for which survey methods might turn out to be most salient include other forms of consumer fraud (such as credit card theft) and experience with public nuisance–type crimes; perception-based measures via the NCVS or supplements might be preferable, on balance, to counts known to be poorly reported to law enforcement.

For several years, BJS has been pursuing several different tracks for overcoming the NCVS's limited capacity for subnational disaggregation. The first is a direct approach: deriving estimates for selected subnational populations consistent with the sample design (the largest of states or metropolitan areas) where there is sufficient sample (pooling multiple years of data as needed), and trying to selectively boost sample size to make more such direct estimates possible. A second approach is indirect: combining existing victimization survey data, existing UCR police-report information, and a variety of community and household statistical information to develop statistical models for victimization of particular types. And a third combines some elements of the other two: effectively defining subnational strata to combine NCVS sample data (and thus develop direct estimates) for groups of places that are similar to each other along particular geographic/demographic lines.

Going forward, to strengthen the survey-based component of the crime data infrastructure, methodological and development work should continue along the lines just described: continuing work on NCVS supplements (or additions to the main survey) on offense types for which survey measures of perceived victimization might have particular weight (including fraud, identity theft, and stalking/harassment) and the refinement of subnational estimation techniques in the NCVS. It is also important that BJS's work on a parallel survey on rape and sexual assault (as recommended by National Research Council, 2014) continue, recognizing the special difficulties of accurate measurement of those sensitive crimes. Another area for ongoing NCVS improvement is to make fuller use of NCVS interviews, including those with households that do not have victimization incidents, to generate data on the community context of crime; this will be discussed further in Section D.3 (Appendix D). The NCVS has to be properly resourced in order to fulfill its core mandate to measure levels and change in crime victimization over time with acceptable levels of precision, much less meet additional demands. The survey approach is too essential a complement to (and a check on) police-report and other crime data to let languish.

New System (Primarily) for "New" Crime: Administrative and External Data

For the "new" offenses in particular, reference to either police report or victimization survey data is likely not the best or most accurate approach to get a sense of levels (both counts and characteristics) and trends in specific offenses. Hence, new systems will be necessary to begin to develop estimates in top-level categories 6–11 in our classification: a crime statistics clearinghouse function to obtain and compile data from external resources including other federal or state agencies. In many cases, this information will be administrative record-type data extracted from the information files of agencies or organizations outside of the police and law enforcement structure. These may include regulatory agencies at the state or federal levels, the courts and other judicial entities, and other public health or welfare agencies. As noted in introducing Section 2.2.3, and consistent with our guidance both here and in Report 1, the objective in marshaling this variety of data is to achieve as much uniformity (across reporting entities) in underlying concept/definition as possible—and as much capacity for disaggregation and reanalysis by attribute/contextual information as is practicable.

This suggestion is more demanding than an ambitious recommendation made by our predecessor National Academies panel to review the full suite of BJS programs. That panel urged BJS to "document and organize the available statistics on forms of crime not covered by" current national crime data series, and to "strive to function as a clearinghouse of justice-related statistical information, including reference to data not directly collected by BJS" (National Research Council, 2009:Rec. 2.1). Our suggestion is not simply to link or refer to external data but to actively assimilate them within national crime statistics. We recognize the task to be difficult and intricate to implement, but necessary to proper understanding of national and subnational crime problems.

Use of alternative data resources is necessary because there are many crime offense types for which neither police-report data nor survey data are apt or workable as a source of offense counts and characteristics. One obvious example would be crimes against government and business that are not specifically spatial in a way that is linked directly to a local police jurisdiction. If a particular convenience store is robbed, the local police will likely be contacted; if a corporation's intellectual property is stolen by hackers attacking from overseas, the local police will not be involved. Indeed, the corporation may not notify any policing agency (local, state, or federal) to avoid reputational damage done by public acknowledgment of its victimization. There are many other reasons that police report and victimization survey data are inherently limited in what they can provide. For example, local law enforcement—traditional police—may not have true jurisdiction relative to other regulatory agencies, such as environmental offenses that would be reportable to state or federal environmental quality authorities. Relevant information on offenses may also be collected by law enforcement agencies that have not typically been thought of as integral to crime reporting (e.g., welfare fraud or other fraud against government agencies, which may be known or reported to state or federal offices of inspector general). For still other offense types, law enforcement officials may not be seen as a relevant or natural point of contact—the quintessential example being fraudulent credit card transactions, in which customers' first and likely only point of contact might be their credit providers.

Conceptually, the challenges of brokering access to alternative, external data resources and processing them loom large—much less marshaling them into a description and assessment of crime offenses from multiple, disparate data sources and providers. Establishing data-sharing arrangements and memoranda of understanding to facilitate them is frequently a difficult chore, even between units of the same federal executive branch agency. Here, the problem is complicated further by seeking information to populate crime statistics from agencies that may view their role as strictly and exclusively civil or regulatory in nature. Entities such as the Securities and Exchange Commission, Federal Trade Commission, Federal Communications Commission, Environmental Protection Agency, U.S. Postal Inspection Service, and numerous inspectors general may—perfectly validly, for their operations—think of their work as cases, complaints, investigations, violations, infractions, or the like, and not really as offenses in the sense of crime. So, to the extent that they see their work relevant to crime statistics, it might be at a very late (resolution) stage—the proffering of criminal charges or formal referral to law enforcement/criminal justice processing. Thus, there may be resistance to sharing or documenting data that strike them as fundamentally preliminary, if related at all, to crime occurrences.

Crime is sufficiently complex that it will require the use of nontraditional data sources and methods in addition to the sources and methods currently in use. The underlying issue is to identify the best way to combine, merge, extract, and analyze administrative-type data from a multitude of agencies and providers, preserving the necessary amount of analytic detail while preserving the privacy of individual persons or other entities represented in those data. Full consideration of these issues is beyond the scope of this panel; indeed, it is the province of the Commission on Evidence-Based Policymaking (2017) mandated by Congress via P.L. 114-140, which issued its final report on September 7, 2017, as well as another National Academies panel tasked to review the use of multiple data sources in federal statistics (National Academies of Sciences, Engineering, and Medicine, 2017a,b).[3]

What is clear is that it is necessary to turn to sources other than police report or victimization survey data to get a more appropriate picture of all crime, and that overcoming the procedural/implementation difficulties will require great effort. To achieve this, the "national" role in "U.S. national crime statistics" will have to be more than a relatively passive aggregator/compiler of data in order to be worthwhile and successful. We return to this point throughout Chapter 3.

2.2.4 Using the Crime Classification as Blueprint

With a next formal conclusion, we return to the issue of crime in crime statistics, the scope of anticipated collection, by emphasizing the importance of maintaining a classification of crime as the foundation for national crime statistics collection.

> **Conclusion 2.4:** The classification of crime for statistical purposes recommended in Report 1 should be used as a blueprint for the nation's crime statistics. It should be used to guide revised measures of crimes currently included in existing data as well as pilot/initial data collection of as-yet-unmeasured crime types—even though it is impossible today for data collection to include every crime in the classification. While it is imperative that the classification undergo

[3]James Lynch, of this panel, was a member of this separate National Academies Panel on Improving Federal Statistics for Policy and Social Science Research Using Multiple Data Sources and State-of-the-Art Estimation Methods, also organized by the Committee on National Statistics.

regular update and revision, initial data collection should begin with this as a base, rather than continue to refine the classification in the absence of data.

Our crime classification attempts to be comprehensive of all crime that occurs, not just that which has traditionally been counted, and it is understood that not every element may be measurable either at the outset or in the near future. It is critically important that the resulting sparseness of the match between our proposed classification, NIBRS, and NCVS be taken as an opportunity for improvement rather than as proof of the impossibility of the task. We considered but ultimately decided against attempting a delineation of highest-priority offenses to measure or, conversely, the ones that will remain aspirational for years. Even if such a comprehensive ranking were possible, we concluded that such a ranking would be inconsistent with one of the basic principles of the classification itself. All of the categories and subcategories have some baseline level of importance. Hence, artificially casting a handful of offense types as most important would seem akin to reprising the UCR SRS's fundamental flaw, putting disproportionate weight on a few select "Part I" crimes and rendering everything else comparatively invisible.

A central observation that follows from interpreting the classification as a framework for understanding the scope of all crime is simply stated but fairly revolutionary for U.S. crime statistics: Any tabulation that wishes to claim to present a comprehensive view of crime in the United States must pay serious attention to categories 6–11 in our classification, commensurate to that given to categories 1–5.

The success of modernizing the crime data infrastructure depends on the extent to which the participating stakeholders accept and endorse the model. Acceptance of the crime classification and the resulting statistics can only occur if the classification is used to generate those statistics. That use must begin somewhere, with trials of prototype measures and phased-in implementation. What is most urgently needed in areas like the various major types of fraud, environmental offenses, or more routinized measures of interference with computer networks is a *start*—the derivation of measures to at least get these crimes and their measures into the same national conversation about crime in the United States as robbery or assault.

The recommended classification of crime should be considered organic and evolving, and should be revised over time to keep up with stakeholder needs and the changing nature of crime itself. There is ample precedent for a structured assessment process, such as those that inform periodic revisions of the International Classification of Diseases or U.S. statistical classifications of industry and occupations. Similarly, we understand that the United Nations Office on Drugs and Crime, whose expert working group developed the International Classification of Crime for Statistical Purposes (ICCS) upon which our proposed classification is patterned, is establishing a standing technical advisory group to assist with implementation and update the ICCS. This is another example that might be considered in formalizing an update process for a U.S. crime classification. But the point we make in the last clause of Conclusion 2.4 is that the classification—like any new structure—needs to be examined and evaluated in practice, through the collection of real data.

The need for reassessment and revision, based on collected data and experience in their analysis, applies to the categories of the classification, its associated attributes, and the specific statistical measures that result from the classification. It is not difficult to find points on which categories or attributes in the proposed classification might be constructed differently. Should hazing be considered separately from other acts that endanger the life of another human (as in category 2.8.1)? Is shoplifting sufficiently different in nature or volume that it should be distinguished from other thefts from businesses or organizations? In the example of ransomware mentioned earlier, is there good reason to carve out a separate category for that particular offense of computer-enabled extortion? In the electronic age, should poor or negligent data stewardship—as in the 2017 data breach involving the credit agency Equifax—be considered a crime of negligence, even though such offenses have typically been limited to negligent acts involving the threat of bodily physical harm? Going beyond individual fine-grained categories, different groups and different stakeholders might reasonably develop completely different classification structures for major offense categories—see, for instance, the extensive taxonomy of fraud developed by Beals et al. (2015) and outlined in Box 5.1 in Report 1. Likewise, in considering the meaning of statistics on things like cybercrime, stolen credit card information, and environmental offenses, analytical paralysis could settle in trying to converge on the ideal measures that should accompany and give meaning to offense counts—whether amount of pollutant released, number of downstream users affected by a computer virus, number of attempts to open credit lines with stolen identity information, or the like. The classification can and should change, and measures associated with it can and should change correspondingly, but that cannot happen productively until initial, prototype numbers are produced and scrutinized.

– 3 –

Coordination and Governance of Modern National Crime Statistics

PROPER CRIME DATA COLLECTION is difficult, but we think that the eventual benefits of a better understanding of crime, improved techniques (developed using those data) for law enforcement to address crime, and increased accountability of and trust in law enforcement entities outweigh the challenges. Moreover, crime results in substantial costs to the nation: Improved measures of crime are imperative for establishing policy priorities across the full range of crime types and for evaluating policy effectiveness in a wide range of areas.

That said, there is a perennial risk that appeals for major change could be considered unrealistic, or simply naive, without grappling with a major underlying question: How can one have confidence in implementing major, new data collections and procedures—on anything approaching a timely basis—when past change has taken so long or proven unsuccessful? The historical record (see Appendix B) abounds with missed opportunities for major change and improvement—the 1958 suggestion that the Federal Bureau of Investigation (FBI) Uniform Crime Reporting (UCR) Program's firm adherence to distinguishing between "Part I" and "Part II" offenses would unduly marginalize the latter and the 1985 call to phase in detailed incident-based crime reporting among them. An obvious further example is that federal law enforcement agencies are mandated by law to report UCR-defined crimes to the UCR Program. This mandate became law in 1988, yet it was not until 2015—following a year-long concerted effort to get a very small set of those federal agencies (including units of the FBI itself) to submit data—that federal submissions warranted publication as a special section in that year's *Crime in the United States* tabulations. This chapter covers our panel's final two formal conclusions. It is about the bigger picture: identifying what our research and our discussions with the wide array of crime data users and stakeholders suggest to be the overriding impediment to change, and suggesting ways to overcome that impediment.

3.1 TAKING OWNERSHIP OF CRIME STATISTICS: DEFINING COORDINATION AND GOVERNANCE FUNCTIONS

In our assessment, the greatest single barrier to change in the nation's crime statistics system is that there is (and has been) no clear, direct answer to the question of who owns—who is ultimately responsible for—the nation's crime statistics, for the full range of crime offenses. Both the FBI and the Bureau of Justice Statistics (BJS) clearly have a strong stake in crime statistics and in understanding and countering crime, generally. But the culture that has evolved over the decades is essentially that the FBI *administers* the UCR program; it *collates* the input data; it *filters* the data through its edit checks and anomaly detection routines. The FBI also *publishes* tabular summaries of the data—historically, as depicted in Section B.4, with strong disclaimers characterizing the national collection as little more than collation, though the FBI has shown more engagement in recent decades. Meanwhile, BJS *supports* the UCR program through its grantmaking authority and its sponsored methodological studies; it *analyzes* the data (along with that from BJS's own National Crime Victimization Survey [NCVS] and other resources) and, through its network of state Statistical Analysis Centers, it *interprets* the data for a wider array of stakeholders. The culture that has evolved over time for national crime statistics collection is one in which no party asserts or has asserted complete program management for data collection on all crime.

In several important ways, the development of this form for U.S. crime statistics collection is completely predictable and understandable:

- A reluctance to impose requirements is directly related to what is certainly a cornerstone of the nation's crime statistics apparatus: the voluntary nature of data contributions to the national/federal level, which itself owes directly to the reservation of most criminal justice and police functions to the states under the U.S. federated system and constitutional government.
- Relatedly, a strong tendency toward decentralization of authority pervades the U.S. justice system. Not only is authority for investigating and combating crime dispersed across thousands of state, local, and tribal agencies, but even enforcement of laws against federal offenses is rarely vested exclusively in a single federal law enforcement agency.

For example, a single offense type like counterfeiting is the investigative/enforcement province of (at least) the U.S. Secret Service, the FBI, and the Federal Trade Commission. Controlled substance–related crimes may be covered (based on their circumstances) by many more agencies ranging from the Drug Enforcement Administration to U.S. Customs and Border Protection to the criminal investigations unit of the Food and Drug Administration.

- The United States is also fairly extreme relative to other countries in the decentralization of its federal statistical system. A great deal of federal statistical activity is vested in 13 principal statistical agencies of the federal government (of which BJS is one), and not a central national statistical office as in some other countries. Moreover, dozens more bureaus and agencies in the federal government devote appreciable resources to conducting statistical activities.

This unique structure is not inherently bad, but the collective consequences are considerable. Deference to the voluntary nature of data contributions means that the federal level cannot directly impose systems or requirements on local law enforcement agencies. Historically, this has suppressed the notion that use of systems or implementation of requirements might be motivated or incentivized, positively or negatively. Put another way, there has been a historical reluctance to apply either "carrots" (e.g., direct financial support) or "sticks" (e.g., linking statistical reporting completeness to eligibility for grant funds) to sway state and local participation in crime statistics. Likewise, dispersed authority for law enforcement has strong virtues. Yet a necessary consequence is that authority for *reporting* or recording those offenses is just as, if not more, dispersed as a result, making it difficult to prioritize measurement of specific offense types or to ensure uniform measurement of concepts.

Often, it has been the case that when new topics of keen policy interest arise and demand quantification, the process of determining what the nation needs to know about the topic and how best to measure it has been done ad hoc. A common approach has been to create new offices or agencies to collect the relevant information or to coerce the new topic into existing data collection structures, more for convenience than accuracy. The confluence of all these decentralizing forces and consequences for crime measurement is probably most strikingly illustrated by the offense of arson. The section on arson in Appendix C gives considerably more detail but, in brief, the United States has no fewer than three designated repositories of national statistics on arson, administratively nested within two separate executive departments—all with explicit legal mandates. Specifically, federal law explicitly authorizes collection of data on arson and criminal misuse of explosives in the FBI, the U.S. Fire Administration of the Federal Emergency Management Agency, and the U.S. Bomb Data Center of the Bureau of Alcohol, Tobacco, Firearms, and Explosives. The three data resources all use slightly differing concepts and rely on voluntary data contributions from different sources of largely unknown accuracy and coverage—and, given the inherent difficulty that may arise in determining whether a fire is deliberately or maliciously set, are all likely inadequate measures of total arson activity in different ways.

In short, because no single entity *owns* or assumes direct responsibility for the nation's crime statistics as a whole, there is neither a strong *coordinating body* nor a defined *governance structure* dedicated to the generation of data on all types of crime. In our assessment, real improvement in national crime statistics hinges critically on the creation and maintenance of both these functions.

3.1.1 Ideal Coordination and Governance Roles

We use both terms, coordination and governance, because the two functions have distinctly different foci and responsibilities. As we will discuss in more detail, delineating the roles separately does not mean that they have to be performed by two different agencies/entities, nor that they necessarily have to be consolidated into one.

Expectations for a strong, national-level coordinating agency for crime statistics require that the agency be actively and directly involved in routine, daily data collection and processing operations. It would bear primary responsibility for regular engagement with source data providers (motivating input to the greatest extent possible while still respecting the ultimately voluntary nature of data contribution to national collections) and managing the resulting data products. Roughly, the key roles of a coordinating body/coordinating agency for crime statistics would include but are not limited to:

- Cultivating an "honest broker" role in interactions with partner agencies, data providers, and providers of computer hardware/software solutions, including negotiation of data-sharing arrangements and memoranda of understanding for data sharing and easing the challenges of implementing necessary changes in data collection content or procedure;
- Maintaining credibility as a source/provider of data in its own right, while providing for analysis (and limited interpretation) of the coordinating agency's own data resources along with those it builds from partner contribution;
- Ensuring and, to the extent possible, enforcing data quality from data provider agencies—deciding what are adequate data as input to the files and what data products are credible for release;
- Managing inflow of data provided by partner agencies and external sources, including providing protections for confidentiality of person, household, business, or organizational records in the data and ensuring

that data quality channels are being followed (e.g., that possible errors detected at the national or state levels are in fact reviewed, revised, and resubmitted); and

- Managing outflow of data and statistical products, including providing feedback and products to source data providers and brokering access to data and summaries to all stakeholders (other agencies, the media, qualified researchers, and so forth).

Again, where the coordination role has tended to be somewhat passive in the past, the great need in a modern system is for top-level, national commitment to the data collection to be continually demonstrated, integrating and assimilating data from a wide variety of sources, and working to ensure their quality. Accordingly, we conclude that this strong national coordination in national crime statistics is essential:

> **Conclusion 3.1:** A stronger federal coordination role is needed in the production of the nation's crime statistics: providing resources for information systems development, working with software providers to implement standards, and shifting some burden of data standardization from respondents to the state and federal levels. The goal of this stronger role is to make crime data collection a product of routine operations.

If coordination is ideally a close-up, hands-on (to the extent possible) role in national crime statistics, then the similarly, critically important role of a governance structure is to take a broader, whole-system view. A core duty of a governance structure is to seek—and act upon the needs of—the broader data provider and stakeholder communities, setting policy for the crime statistics system writ large and providing overall program leadership. Expanding somewhat on these points, the ideal roles of a crime statistics governance structure include but are not limited to:

- Decision-making authority for content (based on stakeholder input), including, in particular, managing a mechanism to maintain and regularly update the underlying classification of crime for statistical purposes and its integral list of attributes;
- Providing voice for (and attention to feedback from) a full range of stakeholders, being responsive to data users' concerns, and involving partner agencies in each others' related data collection efforts;
- Taking a leadership role in promoting data quality, promulgating best practices, and fostering a wider role for analysis of crime data and fuller understanding of crime data among providers and public alike; and
- Facilitating systemwide methodological research and development, including the development and refinement of imputation procedures for

missing data and establishment of protocols for preserving data providers' privacy and the confidentiality of their responses.

In formulating a formal conclusion, on this topic, we emphasize the absence of such a whole-system view at present and urge that such a governance structure—applying to "traditional" and "new" types of crime alike—be established:

> **Conclusion 3.2:** Having an effective governance structure for the complete U.S. crime statistics enterprise is critical. There is currently no entity responsible for reporting on the full range of crimes in the proposed classification (most notably for top-level categories 6–11).

3.1.2 Who Should Coordinate and Govern, and How?

To move forward, it is critical that there be an organizational home for chief coordination and governance roles. In keeping with our panel's charge, we concluded it best not to focus on the exact *who* associated with these roles—whether BJS, FBI, both, or neither—but rather to focus most heavily on the *what* and *how* of crime statistics coordination and governance. To that end, what is of paramount importance is that the coordination and governance of modern crime statistics collection and dissemination be done in accordance with the principles and practices—and the general sensibilities—expected of a statistical agency.

Our panel's administrative parent, the Committee on National Statistics of the National Academies of Sciences, Engineering, and Medicine, published the first edition of its cornerstone document, *Principles and Practices for a Federal Statistical Agency*, in 1992 and later adopted the standard of issuing a revised edition every 4 years. Box 3.1 lists the core principles and practices as expressed in the sixth edition of the report, released earlier in 2017. It is no accident that the ideal coordination and governance roles listed evoke the language and spirit of these principles and practices, in that they (and many other arguments made throughout this report) were made with the tenets of *Principles and Practices* very much in mind. All four principles, and the practices related to them, are particularly vital in the politically charged area of crime statistics, being issues that the existing crime statistics structures have grappled with in the past as well as foreshadowing important developments to come:

- Relevance to policy issues means much more than positing that crime is an important policy topic; it is for adherence to this principle that we emphasize sensitivity to data user and stakeholder feedback, and active engagement with data providers, as strongly as we do.
- Credibility among data users may be the area in which national crime statistics have, historically, fallen shortest, with the overly broad branding

Box 3.1 Principles and Practices for a Federal Statistical Agency

Principles

- Relevance to policy issues: A federal statistical agency must be in a position to provide objective, accurate, and timely information that is relevant to issues of public policy.
- Credibility among data users: A federal statistical agency must have credibility with those who use its data and information.
- Trust among data providers: A federal statistical agency must have the trust of those whose information it obtains.
- Independence from political and other undue external influence: A federal statistical agency must be independent from political and other undue external influence in developing, producing, and disseminating statistics.

Practices

- A clearly defined and well-accepted mission
- Necessary authority to protect independence
- Use of multiple data sources for statistics that meet user needs
- Openness about sources and limitations of the data provided
- Wide dissemination of data
- Cooperation with data users
- Respect for the privacy and autonomy of data providers
- Protection of the confidentiality of data providers' information
- Commitment to quality and professional standards of practice
- An active research program
- Professional advancement of staff
- A strong internal and external evaluation program
- Coordination and collaboration with other statistical agencies

SOURCE: National Academies of Sciences, Engineering, and Medicine (2017c:3,41,45,47,51).

of *Crime in the United States* and insufficient documentation of uncertainty in counts (and completeness of underlying source data) in the UCR police-report data.

- To its credit, adherence to the principle of cultivating trust among data and information providers may be said to be a strong point of the current national crime statistics apparatus. Decades of high and genuinely impressive levels of local-agency voluntary contribution of UCR Summary data is undoubtedly a partial function of crime reporting, but it could not endure without grounding in trust relationships between local agencies, the state UCR coordination network, and the national UCR program. Likewise, BJS's impressive record of methodological innovation in the NCVS is anchored in gaining trust and rapport with survey respondents—to most effectively yet respectfully motivate them to share details about often painful and possibly traumatic experiences. Going forward, modern crime statistics will depend critically on nurturing the same kind of trust relationships with holders of more detailed,

incident-specific information and data from diverse administrative-type data maintainers.
- The principle of independence from political and other undue external influence is a very important one. Consistent with this principle, the true value of national crime statistics is their production in a climate of impartiality and objectivity, working to ensure that definitions and concepts are implemented consistently and uniformly with data quality as the ultimate end objective rather than advancement of any political or administrative agenda.

The practices listed in Box 3.1 are essentially the means by which the principles are met and reinforced; particularly important practices to stress in the development of new crime statistics structures include building/maintaining openness about sources and limitations of the provided data, cooperation with (and responsiveness to) data users, and establishment of an active research program (including methodological development and, substantively, work contributing to the periodic update of the underlying classification and definitions of crime offenses).

Neither of the major agencies that currently collect national crime data, BJS and the FBI, has all of the necessary requirements to fulfill the coordination and governance roles that we envision, though both have some relevant attributes. This observation stems from several things that plain reading of the current statute and regulations justifying the collection of national crime statistics permit stating as basic fact; see Box 3.2 for excerpts of the relevant text. Among these basic facts is that BJS has a very lengthy and detailed list of specific duties etched into its authorizing statute, including a key clause directing the bureau to develop and provide consistent, comparable indicators of crime and victimization. In fact, fair reading of these detailed duties shows very strong correspondence with the ideal coordinating and governance functions we articulated above; legally, BJS has the proper scope and range already built into its legislation, including (respecting the legislative intent when BJS was established in 1979) an obligation to focus on such "new" offense types as corruption and other white-collar offenses. But a significant caveat must be noted: As our predecessor National Research Council (2009) panel reviewing the full suite of BJS programs observed, BJS has never been resourced and funded commensurate with the extensive nature of the duties assigned to it. Then, as now, an overwhelming share of the essentially flat-line funding BJS is allocated is applied directly to conducting the National Crime Victimization Survey (NCVS).

The law and regulation authorizing the UCR Program is decidedly vaguer in letter. As noted in Report 1 and in Box 3.2, it was only 6 decades after the FBI began collecting UCR data that a definition (short of explicit authorization) of the UCR was written into law. That said, the very concise yet very broad

Box 3.2 Current Legal and Regulatory Mandates for the Collection of National Crime Statistics

Federal Bureau of Investigation (FBI)

- **Basic Authorization for Uniform Crime Reporting (UCR) Program:** The Attorney General [shall] acquire, collect, classify, and preserve identification, criminal identification, crime, and other records. *(28 USC § 534(a), as codified by 80 Stat. 616, 1966)*
- *46 Stat. 554 (June 1930)*: There is [hereby] established under the jurisdiction of the Department of Justice a division of the Bureau of Investigation to be known as the Division of Identification and Information; that said division shall be vested with the duty of acquiring, collecting, classifying, and preserving criminal identification and other crime records . . .
- *Current form of 28 CFR § 0.85, formal delegation of Uniform Crime Reporting to the FBI*: The Director [of the FBI] shall: . . .
  - (f) operate a central clearinghouse for police statistics under the Uniform Crime Reporting Program, and a computerized nationwide index of law enforcement information under the National Crime Information Center [and] *[Clause dates to original regulation promulgated in 1963]*
  - (m) carry out the Department [of Justice]'s responsibilities under the Hate Crime Statistics Act *[Clause added in 1990 by rule, 55 FR 28610]*
- *102 Stat. 4468 (1990), first legal definition of UCR Program*: The term 'Uniform Crime Reports' means the reports [authorized under the Attorney General's record collection powers] and administered by the Federal Bureau of Investigation which compiles nationwide criminal statistics for use in law enforcement administration, operation, and management and to assess the nature and type of crime in the United States."

Bureau of Justice Statistics (BJS)

- *42 USC § 3731:* It is the purpose of this subchapter to provide for and encourage the collection and analysis of statistical information concerning crime [and criminal and civil justice system operations] and to support the development of information and statistical systems at the Federal, State, and local levels to improve the efforts of these levels of government to measure and understand the levels of crime . . . The Bureau shall utilize to the maximum extent feasible State governmental organizations and facilities responsible for the collection and analysis of criminal justice data and statistics. *(As originally enacted in 1979 [P.L. 96-157; 93 Stat. 1176], the primary stated interest was information on "crime (including white-collar crime and public corruption)"; however, this parenthetical was dropped when BJS was reauthorized in 1984 [P.L. 98-473; 98 Stat. 2079].)*
- **Designated Duties of BJS,** *selecting those enumerated duties most relevant to national crime statistics (42 USC § 3732(c); unless otherwise noted, clause dates to 93 Stat. 1176, 1979)*: The Bureau [of Justice Statistics] is authorized to[—]
  - (2) collect and analyze information concerning criminal victimization, including crimes against the elderly, and civil disputes
  - (3) collect and analyze data that will serve as a continuous and comparable national social indication of the prevalence, incidence, rates, extent, distribution, and attributes of crime, juvenile delinquency, civil disputes, and other statistical factors related to crime, civil disputes, and juvenile delinquency, in support of national, State, tribal, and local justice policy and decisionmaking *[Amended in 2010 to reference "tribal"]*

Box 3.2 (continued)

- *42 USC § 3732(c), continued*
  - (5) collect and analyze statistical information concerning the prevalence, incidence, rates, extent, distribution, and attributes of crime, and juvenile delinquency, at the Federal, State, tribal, and local levels *[Amended in 2010 to reference "tribal"]*
  - (7) compile, collate, analyze, publish, and disseminate uniform national statistics concerning all aspects of criminal justice and related aspects of civil justice, crime, including crimes against the elderly, juvenile delinquency, criminal offenders, juvenile delinquents, and civil disputes in the various States and in Indian country *[Amended in 2010 to add "and in Indian country"]*
  - (8) recommend national standards for justice statistics and for insuring the reliability and validity of justice statistics supplied pursuant to this chapter
  - (13) provide for the development of justice information systems programs and assistance to the States, Indian tribes, and units of local government relating to collection, analysis, or dissemination of justice statistics *[Clause substantially revised in 1984 from more general "financial and technical assistance" language, and amended in 2010 to reference "Indian tribes"]*
  - (14) develop and maintain a data processing capability to support the collection, aggregation, analysis and dissemination of information on the incidence of crime and the operation of the criminal justice system *[Clause added in 1984]*
  - (16) provide for the collection, compilation, analysis, publication and dissemination of information and statistics about the prevalence, incidence, rates, extent, distribution and attributes of drug offenses, drug related offenses and drug dependent offenders and further provide for the establishment of a national clearinghouse to maintain and update a comprehensive and timely data base on all criminal justice aspects of the drug crisis and to disseminate such information *[Clause added in 1988]*
  - (20) maintain liaison with State, tribal, and local governments and governments of other nations concerning justice statistics *[Amended in 2010 to reference "tribal"]*
- 42 USC § 3732(d): To ensure that all justice statistical collection, analysis, and dissemination is carried out in a coordinated manner, the Director (of BJS) is authorized to—
  - (A) utilize, with their consent, the services, equipment, records, personnel, information, and facilities of other Federal, State, local, and private agencies and instrumentalities with or without reimbursement therefor, and to enter into agreements with such agencies and instrumentalities for purposes of data collection and analysis;
  - (B) confer and cooperate with State, municipal, and other local agencies;
  - (C) request such information, data, and reports from any Federal agency as may be required to carry out the purposes of this chapter;
  - (D) seek the cooperation of the judicial branch of the Federal Government in gathering data from criminal justice records;
  - (E) encourage replication, coordination and sharing among justice agencies regarding information systems, information policy, and data; and
  - (F) confer and cooperate with Federal statistical agencies as needed to carry out the purposes of this subchapter, including by entering into cooperative data sharing agreements in conformity with all laws and regulations applicable to the disclosure and use of data.

Box 3.2 (continued)

- *(42 USC § 3732(e) adds that "Federal agencies requested to furnish information, data, or reports . . . shall provide such information," making cooperation by federal agencies mandatory; both 42 USC § 3732(d) and (e) were part of original 1979 legislation, 93 Stat. 1176)*

NOTE: As of September 1, 2017, the online version of the U.S. Code maintained by the U.S. House of Representatives Office of the Law Revision Counsel was reorganized to assemble a new Title 34, Crime Control and Law Enforcement, from pieces of code that has previously been dispersed in other titles. Accordingly, the citations to Title 42 (The Public Health and Welfare) in this box (and elsewhere in this report) are all now, generally, recodified and renumbered in Title 34. Given the newness of the change, though, and that references have not yet been ported over to print versions of the Code, we retain the older citations in this document.

record collection and maintenance functions vested in the FBI have endured and been generally respected for those many decades. Moreover, the FBI's statutory authority to acquire crime records has been supplemented over time, including being designated by regulation as the Justice Department's collector of hate crime statistics in 1990. The argument that the FBI has cultivated relationships with its "peer" law enforcement agencies at the state and local levels, particularly through its cautious, deliberate Advisory Policy Board (APB) structure, remains a strong one that cannot and should not be upended casually.

In both cases, addressing the crime categories that have long been neglected in national crime statistics will require serious attention to topic domains that may not mesh neatly with the traditional highest priorities of either agency—and there are grounds for concern about the capability of either BJS or the FBI to effectively tap the data resources of external agencies. BJS can legally enter into data collection agreements with all manner of governmental and private agencies. But, practically, there are limits to the amount of suasion that a small bureau, nested within a subunit of the Department of Justice, can be expected to have in negotiations with agencies in other cabinet departments or with pertinent independent agencies such as the Federal Trade Commission, the Environmental Protection Agency, the U.S. Postal Service (and its Postal Inspection Service), and the Securities and Exchange Commission. Work on the "new" crime types will also involve creating interfaces with federal and state Offices of Inspector General (law enforcement agencies but not in the traditional police mold) and a variety of state-level public welfare agencies. BJS might have the legal authority built into its enabling law to broker arrangements with the whole gamut of agencies, but it would likely need additional clout to do so efficiently. The FBI might reasonably be expected to have more clout with some agencies—yet, as observed in opening this chapter, it took nearly 30 years to achieve minimal compliance with explicit, binding, federal law mandate on the FBI's peer federal law enforcement agencies (including the FBI itself) to report UCR-type offenses to the FBI's UCR Program. The FBI's purview has been limited to police-report (or, more properly, law enforcement-report) data, so it would enjoy no special suasion with entities collecting crime-relevant information but who do not view their role as crime investigation. Indeed, some agencies, businesses, or operations might balk at sharing their records and information with a clearly constituted law enforcement agency, even if strictly for statistical purposes.

There are cases that can be made for which agency, or coalition of agencies, should play the strong coordination and governance roles we envision, and we discussed the assignment of the functions extensively in our deliberations. We recognize the legal and regulatory mandates, and general justification, for national crime data collection, and we appreciate the breadth of terrain that must be traversed to begin to come close to a complete picture of crime in the United States. However, we concluded that there is no solid basis for suggesting that any particular "organizational chart" arrangement or designation of duties is superior to any other; we are aware of no relevant research literature or other scientific footing that we could point to in making such a ruling.

Our core argument is that U.S. crime statistics should be coordinated and governed according to the expected sensibilities of a federal statistical agency—which is not the same as saying that a federal statistical agency (and *only* a federal statistical agency, such as BJS) should play those roles. It is difficult to imagine a U.S. crime statistics system without substantial involvement of both BJS and FBI. Yet both would require departures from their status quo operations. BJS is extraordinarily effective within its historically tight resources but would require more resources and likely more legal authority to be successful. And, in support of not only its UCR/NIBRS work but also its maintenance of criminal history records and other data, the FBI should also strongly consider positioning part of its Criminal Justice Information Services (CJIS) Division to directly function as a statistical unit. This change may (perhaps) involve seeking designation and certification as a recognized statistical unit of the federal government pursuant to the Confidential Information Protection and Statistical Efficiency Act (CIPSEA).

Our bottom line is we lack a firm scientific basis for suggesting any particular assignment of strong coordination and governance roles in a new crime statistics infrastructure—and yet we emphatically believe that establishing such roles is critically important, and that there finally be a clear, comprehensible answer to the question, "Who 'owns'/takes responsibility for the nation's crime statistics?" Under the highly decentralized U.S. federal statistical system, such centralized oversight as exists in the system is vested in the U.S. Office of Management and Budget, in the Executive Office of the President. Accordingly, our report's central recommendation is addressed to the single entity that is the most sensible (if not the only realistic) adjudicator of the issue:

> *Recommendation 3.1:* **The U.S. Office of Management and Budget (OMB) should explore the range of coordination and governance processes for the complete U.S. crime statistics enterprise—including the "new" crime categories—and then establish such a structure. The structure must ensure that all of the component functions of generating crime statistics are conducted in concordance with the sensibilities, principles, and practices of a statistical agency. It should provide for user and stakeholder involvement in the process of refining and updating the underlying classification of crime. The new governance process also needs to take responsibility for the dissemination of data products, including the production of a new form of *Crime in the United States* that includes the "new" crime categories.**

It stands to reason that the designation of a coordination and governance structure by OMB should be predicated upon a wide variety of stakeholder input and perspectives from the potential data-providing and data-using agencies, and that statutory language may be necessary to achieve the necessary changes. We will comment further on the final sentence of this recommendation in closing this chapter in Section 3.3.

3.2 RECASTING CRIME DATA COLLECTION AS A FEDERAL-STATE COOPERATIVE PROGRAM

We began this chapter with reference to the federated nature of law enforcement in the United States—the reservation of much police and law enforcement functions to the states and thence to local governments. Accordingly, given the vital role of the states in American law enforcement, another important observation about the coordination of national crime statistics is in order: It is difficult to imagine major change in U.S. crime statistics being successful without national crime data collection being recast as a strong federal-state cooperative or partnership program.

In the UCR's police-report data context, the pivotal role of the states was clear from the outset—if not fully acted upon for a number of years. In the handbook that defined the initial parameters of the UCR Program, the International Association of Chiefs of Police (1929:14) put matters bluntly:

> If crime statistics are ever to be complete for the entire nation, or even for a single state, the state governments must lend a hand. They alone possess the necessary power to enact statutes requiring the submission of uniform crime reports. ... If the states all follow one uniform plan, the

> [U.S.] Department of Justice can then lift bodily the appropriate parts of the state compilations.

As established within the following year, the UCR Program did not entirely follow this guidance, making broader appeals for data submissions to local agencies instead. Still, the very first UCR bulletin invited attention "to the practice prevailing in certain States whereby the State Bureau of identification or other similar agency compiles the crime returns and transmits the reports to this Bureau." The bulletin claimed that "greater accuracy and comprehensiveness will be assured" through this state coordination, and noted that "the results have been encouraging" from the first two states to adopt such a model, Ohio and Utah (Bureau of Investigation, 1930a:1). Thereafter, attention was devoted in the bulletins to new participation by local agencies, and less so about interactions with the states. However, this changed abruptly at the end of the 1960s, at which time more formalized relationships with the states were actively sought. As the 1970 issue of *Crime in the United States* summarized (Federal Bureau of Investigation, 1970:57):

> The last several years the FBI has been actively assisting individual states in the development of statewide programs of police statistics compatible with the national system. These statistical programs have been given impetus by developing statewide computerized law enforcement information systems, of which they are an essential part. Through such mandatory state programs more complete and a better quality reporting is envisioned. Likewise. through coordination with the state agency the data is available for the use of the state, and the collection machinery to the national agency is substantially streamlined.

The report continued by naming specific instances of new state programs for UCR collection—including the strident statement that "in 1969 the FBI seized the collection of all data directly from municipalities and counties in New Jersey and Michigan." Programs were also said to have been established in 1969–1970 in Kentucky, Rhode Island, Pennsylvania, Minnesota, Nebraska, Florida, and Wisconsin, with some partial implementation in California as well. In the decades that have followed since then, what has evolved in the national UCR program is the development of a network of state UCR programs—through which virtually all local-agency-submitted reports are channeled before forwarding on to the FBI and national collection. (Exceptions to reporting through the state are now rare, typically coinciding with technical difficulties at the state level or—as is currently the case for the state of Mississippi—lack of a recognized state program.) Collectively, this network of state agencies has a support arm/service organization of its own, in the Association of State Uniform Crime Reporting Programs (ASUCRP).

3.2.1 Observations from State Crime Reporting Laws and Regulations

In Appendix E, we compile extracts of crime reporting laws and regulations defined in state legal and administrative codes (as distinct, whenever possible, from related law on the maintenance and use of criminal history records, for instance); the appendix also summarizes the status of incident-based crime reporting in each state. The summaries of state law are admittedly somewhat cursory in nature, being limited to immediately obvious sections of code and support documentation, and not an exhaustive canvass of every state's legal documents. Even at that, the material in the appendix is both dense and difficult—but impressions from it are absolutely essential to understanding some features of current crime data collection and the prospects for future approaches.

Above all, reference to the language of state laws makes abundantly clear that partnership with the states is not simply desirable—if only to reduce the number of data input streams from almost 20,000 individual law enforcement agencies to roughly 50—but virtually essential. Throughout this report, we have tried to take care to characterize the voluntary nature of local agency crime reporting as effectively being voluntary *to the national program*. But the same is not true of reporting *from local agencies to the state*. By our reading and count in Appendix E, the local law enforcement agencies operating within 43 states are legally obligated to submit crime report data to a state-level entity—typically, the state police/highway patrol, justice department, or other entity that is designated as the state UCR program/coordinator. An additional three states have legal code that suggests "cooperation" with the national UCR program but does not explicitly require transfer of data to the national level; four states' laws appear to leave the issue unresolved. By comparison, it is hardly unprecedented for local agencies to be explicitly required by state law to contribute data to the national UCR program—eighteen are so required, and another seven are encouraged to contribute to the national collection without it being explicitly mandated (the other half of state codes appear to be silent on the issue).[1]

Another key observation from the state law appendix is the very encouraging signals that come from some of the most recent revisions of state crime reporting laws to update—signs of intent to not only upgrade to more detailed incident-based crime reporting, but also to do so on a quicker pace than the national program. As we will discuss further in Section D.1, one of the critical barriers to rapid adoption of the National Incident-Based Reporting System (NIBRS) by states and localities was the lack of early NIBRS-format adoption by large, major jurisdictions whose commitment to the new reporting standard would serve as a motivational example to other jurisdictions. Accordingly, the fact that both California and Texas have enacted laws committing to

---

[1]Indeed, it is a curious quirk in the letter of Ohio state law that appears to put higher priority on reporting crime to the national UCR program than to the state coordinator.

replace UCR Summary submissions with NIBRS format *ahead* of the FBI's 2021 sunset date is remarkable—Texas even more so given its longstanding previous reputation as having no plans for NIBRS implementation whatsoever. Indiana has also required all local agencies to report in NIBRS format by mid-2018. Illinois's new crime reporting statute hints at conversion statewide, Chicago having long been a holdout and having made substantial investment in overhauling its internal information management systems.

Going further along these same lines, our review of the state language does not purport to be exhaustive; it may miss references to information and data requests that are peppered throughout the definitions of offenses in state law, for instance. Still, we list in the appendix offense types that are identified as particularly high priorities for crime measurement when those are apparent in state reporting law. Domestic violence, drug-related offenses, computer-involved crime, human trafficking, and other specific offenses are all issues that one or more states have legally insisted be addressed in the state's crime reporting program. These and other emerging examples—perhaps most notably New York's statutory call for environmental offenses and insurance fraud to be treated commensurately with other crime types in the state UCR program's annual reports—suggest that state-level interest in the substance of modernized crime statistics (and the "new" crime types) in particular may be encouraging grounds for more active partnership work.

In Section D.2, we will discuss in greater detail the notion of data quality as the primary "value added" to local law enforcement stakeholders (and data users)—as the thing that makes it beneficial to local agencies' routine operations and thus motivates cooperation with the national crime statistics program. Along this line, the review of state legislative language is also useful to consider; data quality assurances may resonate well with states like Georgia, Illinois, and South Carolina, which have particularly strong record-audit or quality-control-inspection provisions built into their state codes.[2]

Two other comments are motivated indirectly by the summaries of state law in Appendix E. First, an observation based on what is *not* covered by the appendix is that similar brokering arrangements with state-level coordinators might be appropriate to manage data inputs from relevant state, local, or private authorities, if those records aren't immediately or already directly shared with some federal counterpart. For instance, state environmental quality or consumer protection agencies may be important sources of records on environmental and consumer fraud offenses, drawing their information from local/municipal or other sources. In some cases, these agencies may be legally required to report to a state-level entity in much the same way as local police are

[2]It is useful to see some other examples spelled out in state code, including Connecticut's provision that makes the FBI (and the national UCR program) the arbiter of whether crime data submitted to the state are incomplete or incorrect.

commonly obligated to report UCR data to the state. Similarly, there are some offenses such as fraud against the government (e.g., fraudulent medical benefit claims) in which state attorneys general or inspectors general may be the only potential, viable source—likewise for highly state-differing policies on gambling or alcohol enforcement.

The difficulty of parsing the language of state law occasions the second comment, which is that the construction and maintenance of crosswalks between offense codes in different data systems are essential functions. Our proposed classification uses behavior-based definitions as a workaround to the intricate detail in state criminal code language; we define offenses based on the behavior and not on trying to find some lowest-common-denominator wording across dozens of state criminal codes. We believe in this approach, with an ideal vision for crime statistics being one in which data providers (including police officers generating reports or survey respondents answering questions) never have to go through the cognitive process of recalling codes and deciding which classification labels are most appropriate to the incident at hand. That said: To be useful in daily operations, every state and local law enforcement information management system has to be able to handle and work with offense codes, based on state (and local) law, for those offenses for which officers have arrest power and departments have booking power. Difficult as the task of finding and referencing passages of state law related to crime reporting is, it pales in comparison to the task of building crosswalk files between offense codes based on state law and different coding systems (e.g., NIBRS codes or our classification's codes, plus associate attributes)—and keeping those crosswalk files up to date with enactment of new law. Building these crosswalk files exclusively at the federal level and filtering them down would be difficult and likely fraught with error. Building and maintaining them is also not something that should be imposed directly on information management system vendors/solution providers, though they will obviously be essential to incorporating them into the systems. Crosswalk construction and maintenance may be a very daunting task, but one that is only really tractable through partnership with the states.

3.2.2 Examples of Federal-State Cooperative Programs in the Statistical System

The coordination of data transfers between multiple federal agencies is a difficult undertaking, much less the management of data inputs from a much wider variety of state, local, and private entities. This is arguably an even harder task when data contribution and sharing is premised on said contributions being voluntary in nature. We have argued that treating the general problem as a federal-state cooperative arrangement is virtually essential to making a new system work. But it follows that there may be lessons that might be learned from other statistical agencies in other policy domains, in which federal-state compacts are fairly frequent. We believe it appropriate to mention a few illustrative programs here that might be useful to consider. In doing so, we reiterate that we know of no example federal-state cooperative that is a direct, obvious, interchangeable match to the problems of crime data collection—nor do we know of any federal-state arrangement that has uniquely solved all the myriad problems associated with data transfers—but merely that they may provide useful perspectives for framing implementation and methodological work.

A major example of a federal-state cooperative data collection program that might be a useful model to consider is the Bureau of Labor Statistics (BLS) Survey of Occupational Injuries and Illnesses (SOII).[3] Businesses and establishments are required to keep records of work-related injuries and illnesses by Occupational Safety and Health Administration (OSHA) regulations, and the SOII program has served as a means of regularly compiling information from these records. Each year, a sample of employers is selected, and is requested to submit a mix of summary-count data and incident-specific detail based on their in-house OSHA records—the summary-count data applying to all workplace injuries and illnesses and the incident-level detail being requested about those cases that lead to one or more days of missed work. The responses are channeled through BLS's network of state coordinators. One possible weakness of the SOII model for crime data collection is the frequency of collection—the data are requested once a year, in January, rather than on a monthly (more continuous basis) throughout the year. But what makes the SOII model most useful as an illustrative example is that BLS (together with its state partners) assumes the burden of classification and standardization of responses from the responding employers. In the survey's data collection forms, critical variables including the injured worker's actual occupation (or job title) and the nature of the injury/illness are free-text responses; the reporting business or establishment is not asked to internalize a set of classification rules and self-apply them. Instead, BLS and its partners assume the responsibility for working with reported job title and other contextual information and assigning the proper code from the U.S. Standard Occupational Classification and standardizing injury type, among others. It is a time-intensive process—businesses are alerted to their selection in the SOII sample and asked for data in January (being asked to provide data for the previous calendar year), and estimates are typically produced in October or November. Hence, the survey's ongoing work to carry out these tasks makes it a useful case to study; the

[3]Additional information on the survey is available at http://www.bls.gov/iif and https://www.bls.gov/respondents/iif/faqs.htm. The most recent data collection instruments for the survey, and additional descriptive information, is included as part of BLS's filing for clearance by the U.S. Office of Management and Budget to conduct the survey, available at https://www.reginfo.gov/public/do/PRAViewICR?ref_nbr=201706-1220-005.

agency is constantly trying to improve computer-based autocoding of the free-text responses to make the program more efficient (see, e.g., Measure, 2014). To the extent that BLS is able to directly obtain administrative record-type data in lieu of filled survey forms—a more direct analogue to receipt of local law enforcement records rather than summary "reports"—its work to use such data in the SOII will be useful to monitor as well.

A second, longstanding federal-state cooperative program is one that is arguably closest in content to some crime data: the nation's core collection of vital statistics. The National Vital Statistics System is the oldest and most successful example of intergovernmental data sharing in public health and is a potentially instructive example to consider because it faces some of the same structural problems as faced in crime statistics. Information from vital records are provided through contracts between the National Center for Health Statistics (NCHS) and vital registration systems operating in the 57 vital registration jurisdictions (the 50 states, New York City, the District of Columbia, and the territories) that are legally responsible for the registration of births, deaths, fetal deaths, marriages, and divorces. Unit records with information on births, deaths, and fetal deaths are, for the most part, sent to the registration jurisdiction using electronic systems from hospitals and funeral directors. They are then sent electronically from the jurisdictions to NCHS. The development of model state laws, standard certificates, and standardized methods of data collection and processing support the creation of national data files. While automating the registration of vital events increased the timeliness and quality of the data, the adoption of different hardware and software systems by the jurisdictions has made it much more complicated and expensive to change the content of the data collection. Work is underway to address the challenges of maintaining and improving electronic systems for collecting and transferring information.

A final example to consider for its organizational aspects is the dense network of regional and state field offices maintained by the National Agricultural Statistics Service (NASS). A single "state" office covers several of the New England states; otherwise, every state has a NASS state field presence. The day-to-day function of NASS field staff is to work sufficiently in concert with state agricultural departments—collecting data and items on a vast array of crop- and practice-specific collections—that they are practically embedded in the state agencies. Moreover, the National Association of State Departments of Agriculture (NASDA) that functions as an organizational arm for the state agriculture agencies has also served as a corps of field representatives and interviewers for NASS censuses and surveys.

3.2.3 State Statistical Analysis Centers (SACs)

A final note that must be made in this context is that, in the current crime statistics system, the states have had two fairly well-defined roles. The first, the state UCR programs, has already been discussed and should be expanded upon in new collection strategies. But it is also important that the second be continued as well—the network of Statistical Analysis Centers (SACs) that receive some support funding through BJS grant monies and that have long served a critical role as interpreters and analysts of criminal justice system data, including both BJS and FBI/UCR data. The SACs publish reports and research results for the public, the media, and the broader research community, but their primary (and crucial) constituency is state legislative and executive decision makers. The National Crime Statistics Exchange (NCS-X) initiative will provide useful insights that can be used to demonstrate the utility of incident-based data and encourage still further NIBRS participation. The SACs are likely better poised to perform these kinds of analyses than many individual police departments, and so their role in examining the first full NIBRS data will be vital.

3.3 DATA REPORTING AND DISSEMINATION

Managing and enhancing the outflow of data and data products is an important part of the ideal, beginning-to-end coordination and governance roles we envision for national crime statistics. Issues related to publishing annual reports and other tabulations are crucial to the credibility of the data collection enterprise as a whole; this is an area, after all, in which the annual publication of a volume entitled *Crime in the United States* has constrained the scope of what is commonly viewed as crime and given an undue impression of quality and completeness. Hence, the final sentence of Recommendation 3.1, on data products and their dissemination, involves issues that deserve serious attention.

Weighing all that we have heard from data users and stakeholders, we conclude that new crime data products—and data dissemination more generally—must necessarily progress along two lines. In our assessment, there remains a necessary role for an annual, comprehensive report that reflects all crime types—yet the issuance of data and analyses should not be inextricably linked to the preparation cycle for the full-fledged annual report. Hence, the aim for crime data products should extend beyond a printed volume, and instead aim for more dynamic presentation and availability of a crime data repository that is updated on a flow basis.

There is a certain inevitability to an annual report as a partial dissemination vehicle for crime statistics. Per our examination of state crime reporting laws in Appendix E, 34 of 50 states are explicitly required to issue a crime statistics report with some frequency, typically annually, to state officials (executive or legislative) or to the general public; though law and regulation appear silent about production of a report in the remaining states, several of them do issue annual reports nonetheless. At the national level, both *Crime in the United States* (UCR) and *Criminal Victimization* (NCVS) have been annual fixtures for decades. Issuance of the annual report is one way (though certainly not the only way) to construct time series and permit annual benchmarking/comparisons.

But the indispensable role of a comprehensive annual crime report, for purposes of nurturing a new crime statistics system, is simply that what is most necessary for the "new" and heretofore neglected-in-national-statistics offense types in our proposed classification is getting them into the same national conversation about crime as the "traditional," index, violent and property crimes. Accordingly, the best and most logical way for that to happen is for the new and traditional offenses to be presented and discussed, with commensurate attention, in the same forum. For that purpose, an annual report that strives for some synthesis rather than simple presentation of counts is ideal. No data user or stakeholder is well served by a static, incomprehensible mass of tables published solely for publication's sake. In contrast, a report that captures some indicators of how, when, and why both "traditional" and "new" crimes occur is beneficial, even at the expense of some more preparation time.[4]

Dissemination of modern crime data should also be paired with modern, dynamic data access tools, capable of handling a broad range of user data queries. The important part in introducing the two tracks for development of dissemination approaches is decoupling them—not tying release of data through online means solely to the production cycle of an annual report. Hence, an online crime data repository and access tool should be updated as often as practicable. It should be capable of generating standard tabulations, but do so in a way that conveys the degree of uncertainty and types of error associated with the estimates. It should also facilitate researchers' access to data files and associated metadata, permitting intensive data users to conduct their own analyses offline. Ideally, it could also permit user-constructed data visualizations and graphics. These are all lofty objectives—it is easy to hope for a high-quality, national-level, at- or near-real-time dashboard, but that is truly a distant wish. As the range and content of national crime statistics grow, from an NCS-X–supplemented national NIBRS file to the fuller, multisource collection enabled by and based upon a new classification of crime, there will inevitably be a learning curve. Challenges will be encountered and iterative improvements

[4]An admitted wild card in consideration of annual reporting is that a comprehensive annual report should, necessarily, combine insights from police-report, survey, and administrative-type/external sources in tandem—yet the UCR-based *Crime in the United States* and the NCVS-based *Criminal Victimization* have traditionally had staggered release dates. Over the years, the temporal separation between the reports has created the potential for confusion or mixed messages. However, whether it is possible to more directly synchronize the releases (or whether it is worth doing so) is an open question that we did not attempt to resolve.

made, but the point is to place prototypes (annual report and data access tool alike) into production and improve from there.

Finally, as the traditional UCR Summary is phased out in 2021, it is important that the annual dissemination of information that has been routinely presented in *Crime in the United States* continue to allow for the possibility of assessing changes over time in rates of traditionally measured crimes in ways that preserve the continuity of the statistical series. Variation across agencies and states in full NIBRS adoption should be anticipated and planned for, and information about changes in agency participation and data completeness should be fully documented so that sources of potential errors in crimes such as homicide are transparent and understood by the public.

3.4 REMAINING IMPLEMENTATION AND METHODOLOGY ISSUES

Appendix D includes longer-form discussions of some additional important topics related to methodology and implementation, mentioned in brief in the main text but fuller consideration of which might otherwise disrupt the narrative flow. These include:

- Examining the commonly cited barriers that impeded adoption of NIBRS for 3 decades, accompanied by discussion of what has changed now to make success more likely;
- Arguing that improved data quality should be the major goal of voluntary contribution of data to national crime statistics;
- Observing that—rather than warn against overly simplistic comparisons of crime rates across jurisdictions and time—the dissemination platform for modern crime statistics should also provide easy access to the kind of contextual and community data that enables effective comparison; and
- Describing the challenges of getting a better reading on crime involving businesses and organizations as actors.

3.5 CONCLUSION

The path ahead for U.S. national crime statistics is neither easy, inexpensive, nor short in terms of time to completion. If BJS and the FBI are successful in getting a designed sample of NIBRS-responding agencies from which sound inference can be drawn to a larger whole—and if the FBI is successful in its goal in sunsetting the UCR Summary Reporting System by 2021 in favor of full NIBRS—the fact will remain that it took more than 35 years to reach that goal. To modernize crime statistics and to equip the nation to better understand problems of crime and victimization will require a large investment in collecting data on issues that are currently all but ignored in current crime measurement efforts. An adequately resourced NCVS and a full-participation (and fully representative) NIBRS are important steps along the way, but real benefit will arise from finally starting to shed light on too-long neglected or obscured offenses, consistent with a rigorous classification.

In this report, we have suggested adding a third core crime data component: administrative-records-type data and information on regulatory violations, amassed through a crime statistics clearinghouse function. Constructing this clearinghouse is a difficult but essential task. However, in closing this report, we would be remiss if we did not at least mention an important point along these lines. Some particularly difficult-to-measure offenses (e.g., trafficking in persons or migration offenses) are very likely to elude detection in third-party, administrative-type data resources, just as many traditional offenses elude detection in the police-report, UCR/NIBRS data. In the same way that survey data (i.e., the NCVS) play an indispensable role in providing an estimate of crime not reported to police, there is a necessary role for alternative methodology for validating counts or characteristics of relevant offenses that are not reported in regulatory data. A modern crime data infrastructure should enable long-term research and development on innovative technologies and methodology for measuring crime just as it should enable such research on the utility of specific metrics, as in our Conclusion 2.2 on the harms or cost of crime. We have focused in this report on the three, major, essential pieces of a new crime data infrastructure—already a difficult goal to attain—but the longer-term research interests are vital as well.

Attribute-driven and incident-based reporting using the classification proposed by our panel means upending concepts that have been at the heart of crime data collection for decades—not least of which are a shift away from defining crime by the strict language of state penal code violations and the treatment of some offenses that carry administrative/regulatory sanctions in the same manner as those punishable by corrections. It also challenges crime data providers, collectors, aggregators, processors, and users alike to embrace some markedly different approaches—the most fundamental, and simple, being the parsing of multiple offense codes for simple incidents, relative to the cognitive oversimplification of the traditional UCR Summary that collapses each incident to a single offense. Full implementation will mean major changes in data collection forms, software, and processes, all of which take time and commitment. Yet such radical transformation becomes justified if the principles on which this classification and subsequent data structures are borne out, providing a much more robust understanding of the nature of crime and firmer ground for developing policies to combat it.

References

Addington, L. A. (2007). Using NIBRS to study methodological sources of divergence between the UCR and NCVS. In J. P. Lynch and L. A. Addington (Eds.), *Understanding Crime Statistics: Revisiting the Divergence of the NCVS and UCR*, Cambridge Studies in Criminology, Chapter 8. New York: Cambridge University Press.

Akiyama, Y. and S. K. Propheter (2005, April). *Methods of Data Quality Control: For Uniform Crime Reporting Programs.* Washington, DC: Federal Bureau of Investigation.

Alabama Law Enforcement Agency (2016). *Crime in Alabama—2015.* Montgomery: Alabama Law Enforcement Agency.

Apel, T. (2015, October 1). Miss. crime stats revealing. *Jackson Clarion-Ledger.* Published online: http://www.clarionledger.com/story/news/2015/10/01/fbi-miss-capital-city-homicides/73181106/.

Arizona Department of Public Safety (2016). *Crime in Arizona 2015.* Phoenix: Access Integrity Unit, Arizona Department of Public Safety.

Bauer, L. (2004, September). *Local Law Enforcement Block Grant Program, 1996–2004.* NCJ 203096. Washington, DC: U.S. Department of Justice, Bureau of Justice Statistics.

Beals, M., M. DiLiema, and M. Deevy (2015, July). *Framework for a Taxonomy of Fraud.* Joint collaboration of Financial Fraud Research Center at the Stanford Center on Longevity and the FINRA Investor Education Foundation. Stanford, CA: Stanford Center on Longevity.

Beck, A. (2016, August). *Beyond Shrinkage: Introducing Total Retail Loss.* Commissioned report. Arlington, VA: Retail Industry Leaders Association.

Bureau of Investigation (1930a, August). *Uniform Crime Reports for the United States and Its Possessions: Monthly Bulletin for August, 1930*, Volume 1(1). Washington, DC: Government Printing Office.

Bureau of Investigation (1930b, October). *Uniform Crime Reports for the United States and Its Possessions: Monthly Bulletin for October, 1930*, Volume 1(3). Washington, DC: Government Printing Office.

Bureau of Investigation (1931a, June). *Uniform Crime Reports for the United States and Its Possessions: Monthly Bulletin for June, 1931*, Volume 2(6). Washington, DC: Government Printing Office.

Bureau of Investigation (1931b, May). *Uniform Crime Reports for the United States and Its Possessions: Monthly Bulletin for May, 1931*, Volume 2(5). Washington, DC: Government Printing Office.

Bureau of Investigation (1932, April). *Uniform Crime Reports for the United States and Its Possessions: First Quarterly Bulletin, 1932*, Volume 3(1). Washington, DC: Government Printing Office.

Bureau of Investigation (1933, April). *Uniform Crime Reports for the United States and Its Possessions: First Quarterly Bulletin, 1933*, Volume 4(1). Washington, DC: Government Printing Office.

Bureau of Investigation (1935, April). *Uniform Crime Reports for the United States and Its Possessions: First Quarterly Bulletin, 1935*, Volume 6(1). Washington, DC: Government Printing Office.

California Department of Justice (2014, April). *Criminal Statistics Reporting Requirements*. Sacramento: Bureau of Criminal Information and Analysis, Criminal Justice Statistics Center, California Department of Justice.

Central Statistics Office (2015, June). *Central Statistics Office: Review of the Quality of Crime Statistics*. Cork, Ireland: Government of Ireland, Central Statistics Office.

Chilton, R. and W. C. Regoeczi (2007). Impact of employment, family structure, and income on NIBRS offense, victim, offender, and arrest rates. *Justice Research and Policy* 9(2), 9–29.

Commission on Evidence-Based Policymaking (2017, September 7). *The Promise of Evidence-Based Policymaking: Report of the Commission on Evidence-Based Policymaking*. Washington, DC: Commission on Evidence-Based Policymaking.

Criminal Justice Information Services Division (2015). *UCR Program Quarterly, January 2015*. Number 1. Clarksburg, WV: Federal Bureau of Investigation.

Criminal Justice Information Services Division (2016a). *UCR Program Quarterly, April 2016*. Number 2. Clarksburg, WV: Federal Bureau of Investigation.

Criminal Justice Information Services Division (2016b). *UCR Program Quarterly, July 2016*. Number 3. Clarksburg, WV: Federal Bureau of Investigation.

Criminal Justice Information Services Division, Virginia State Police (2011, July). *Virginia State Police Uniform Crime Reporting incident-based procedure guide manual.* Richmond: Department of State Police.

Delaware Criminal Justice Council, Statistical Analysis Center (2016, August). *Crime in Delaware 2011–2015: An Analysis of Serious Crime in Delaware.* Prepared in conjunction with the State Bureau of Identification. Dover: Delaware Criminal Justice Council.

Division of Investigation (1934a, April). *Uniform Crime Reports for the United States and Its Possessions: First Quarterly Bulletin, 1934*, Volume 5(1). Washington, DC: Government Printing Office.

Division of Investigation (1934b, July). *Uniform Crime Reports for the United States and Its Possessions: Second Quarterly Bulletin, 1934*, Volume 5(2). Washington, DC: Government Printing Office.

Ennis, P. H. (1967, May). *Criminal Victimization in the United States: A Report of a National Survey.* Prepared by National Opinion Research Center and submitted to the President's Commission on Law Enforcement and Administration of Justice. Washington, DC: U.S. Government Printing Office.

Federal Bureau of Investigation (1935a, July). *Uniform Crime Reports for the United States and Its Possessions: Second Quarterly Bulletin, 1935*, Volume 6(2). Washington, DC: Government Printing Office.

Federal Bureau of Investigation (1935b, October). *Uniform Crime Reports for the United States and Its Possessions: Third Quarterly Bulletin, 1935*, Volume 6(3). Washington, DC: Government Printing Office.

Federal Bureau of Investigation (1939). *Ten Years of Uniform Crime Reporting, 1930–1939: A Report by the Federal Bureau of Investigation.* Washington, DC: U.S. Department of Justice.

Federal Bureau of Investigation (1940a, April). *Uniform Crime Reports for the United States and Its Possessions: First Quarterly Bulletin, 1940*, Volume 11(1). Washington, DC: Government Printing Office.

Federal Bureau of Investigation (1940b, July). *Uniform Crime Reports for the United States and Its Possessions: Second Quarterly Bulletin, 1940*, Volume 11(2). Washington, DC: Government Printing Office.

Federal Bureau of Investigation (1950, April). *Uniform Crime Reports for the United States and Its Possessions: Semiannual Bulletin, 1950*, Volume 50(1). Washington, DC: Government Printing Office.

Federal Bureau of Investigation (1953, January). *Uniform Crime Reports for the United States and Its Possessions: Annual Bulletin, 1952*, Volume 23(2). Washington, DC: Government Printing Office.

Federal Bureau of Investigation (1954, January). *Uniform Crime Reports for the United States and Its Possessions: Annual Bulletin, 1953*, Volume 24(2). Washington, DC: Government Printing Office.

Federal Bureau of Investigation (1958). *Uniform Crime Reports for the United States: Special Issue, 1958.* Washington, DC: Government Printing Office.

Federal Bureau of Investigation (1960). *Crime in the United States: Uniform Crime Reports—1960.* Washington, DC: U.S. Department of Justice.

Federal Bureau of Investigation (1961). *Crime in the United States: Uniform Crime Reports—1961.* Washington, DC: U.S. Department of Justice.

Federal Bureau of Investigation (1970). *Crime in the United States: Uniform Crime Reports—1970.* Washington, DC: U.S. Department of Justice.

Federal Bureau of Investigation (1992). *Uniform Crime Reporting Handbook: National Incident-Based Reporting System Edition.* Washington, DC: U.S. Department of Justice, Bureau of Justice Statistics.

Federal Bureau of Investigation (2004). *UCR: Uniform Crime Reporting Handbook.* Washington, DC: U.S. Department of Justice, Bureau of Justice Statistics.

Federal Bureau of Investigation (2012). *Uniform Crime Reporting Program Data: Supplementary Homicide Reports, 2012.* Codebook. ICPSR 35023. Ann Arbor, MI: Inter-university Consortium for Political and Social Research [producer and distributor].

Federal Bureau of Investigation (2013). *Criminal Justice Information Services (CJIS) Division Uniform Crime Reporting (UCR) Program—Summary Reporting System (SRS) User Manual.* Version 1.0. Washington, DC: U.S. Department of Justice, Federal Bureau of Investigation.

Federal Bureau of Investigation (2015, October 13). *Criminal Justice Information Services (CJIS) Division Uniform Crime Reporting (UCR) Program—National Incident-Based Reporting System User Manual.* Version 2.0. Washington, DC: U.S. Department of Justice, Federal Bureau of Investigation.

Flatley, J. and J. Bradley (2013, January 24). *Analysis of variation in crime trends: A study of trends in 'comparable crime' categories between the Crime Survey for England and Wales and the police recorded crime series between 1981 and 2011/12.* Methodological note. London: Office for National Statistics.

Garda Síochána Inspectorate (2014, October). *Crime Investigation Report.* Dublin: Garda Síochána Inspectorate.

Greenfield, V. A. and L. Paoli (2013). A framework to assess the harms of crime. *British Journal of Criminology 53*, 864–885.

Haas, S. M., J. P. Jarvis, E. Jefferis, and E. Turley (2007). Gun availability and crime in West Virginia: An examination of NIBRS data. *Justice Research and Policy 9*(2), 139–159.

Harrell, E. and L. Langton (2013, December). *Victims of Identity Theft, 2012.* NCJ 243779. Washington, DC: U.S. Department of Justice, Bureau of Justice Statistics.

Her Majesty's Inspectorate of Constabulary (2012). *The crime scene: A review of police crime and incident reports.* London: Her Majesty's Inspectorate of Constabulary.

Hickman, M. J. (2005, July). *Justice Assistance Grant (JAG) Program, 2005.* NCJ 209333. Washington, DC: U.S. Department of Justice, Bureau of Justice Statistics.

Humpherson, E. (2016, September 27). *Assessment of statistics on recorded crime in Scotland.* Letter of confirmation from Director General for Regulation, United Kingdom Statistics Authority, to Roger Halliday, Chief Statisician and Head of Performance, Scottish Government. London: United Kingdom Statistics Authority.

Idaho State Police (2017, July 1). *Crime in Idaho 2016.* Meridian: Bureau of Criminal Identification, Idaho State Police.

Illinois State Police (2016). *Crime in Illinois 2015.* Springfield: Illinois State Police. Published online in segments at http://www.isp.state.il.us/crime/cii2015.cfm.

International Association of Chiefs of Police (1929). *Uniform Crime Reporting: A Complete Manual for Police; Report of the Committee on Uniform Crime Records.* New York: J.J. Little and Ives Company.

Kansas Bureau of Investigation (2012, January). *Kansas Incident Based Reporting System Handbook* (Sixth ed.). Topeka: Kansas Bureau of Investigation.

Kentucky State Police (2014). *Crime in Kentucky—2013.* Frankfort, KY: Kentucky State Police.

LaFree, G. (forthcoming). The evolution of terrorism event databases. In E. Chenoweth, A. Gofas, R. English, and S. Kalyvas (Eds.), *Handbook on Terrorism.* New York: Oxford University Press.

LaFree, G., L. Dugan, and E. Miller (2015). *Putting Terrorism in Context: Lessons from the Global Terrorism Database.* London: Routledge.

Law Enforcement Support Division, Texas Department of Public Safety (2015, February 3). *DPS Texas Incident Based Reporting System (TIBRS) Training and Reference Manual.* Austin: Texas Department of Public Safety.

Louisiana Uniform Crime Reporting Program (2016, November 1). *Crime in Louisiana 2014.* Joint project of the Louisiana Commission on Law Enforcement and Administration of Criminal Justice and the Louisiana Sheriffs' Association. Baton Rouge.

Maine State Police (2016, September). *Crime in Maine 2015.* Augusta: Maine Department of Public Safety.

Maltz, M. D. (1999, September). *Bridging Gaps in Police Crime Data: A Discussion Paper from the BJS Fellows Program.* Bureau of Justice Statistics and Federal Bureau of Investigation Uniform Crime Reporting Program. NCJ 176365. Washington, DC: U.S. Department of Justice, Bureau of Justice Statistics.

Maxfield, M. G. (1999). The National Incident-Based Reporting System: Research and policy applications. *Journal of Quantitative Criminology 15*(2), 119–149.

McClure, C. L. (2016). *Felony Level Sex Offenses, 2015: Crime in Alaska Supplemental Report.* Anchorage: Alaska Department of Public Safety.

Measure, A. C. (2014). Automated coding of worker injury narratives. In *Proceedings of the Government Statistics Section*, Paper prepared for the Joint Statistical Meetings. Alexandria, VA. American Statistical Association.

Michigan Incident Crime Reporting Unit (2016, May). *Michigan Incident Crime Reporting (MICR) Specification Handbook.* Revision of January 1998 document. Lansing: Michigan Department of State Police.

Miller, E. (2014). *Patterns of Terrorism in the United States, 1970–2014.* College Park, MD: National Consortium for the Study of Terrorism and Responses to Terrorism (START).

Montana Board of Crime Control (2016, August). *Crime in Montana: 2014–2015 Report.* Helena: Statistical Analysis Center, Montana Board of Crime Control.

Montana Board of Crime Control, Statistical Analysis Center (2015, January). *Montana Incident-Based Reporting System (MTIBRS) Standards Handbook.* Helena: Montana Board of Crime Control.

Municipality of Anchorage (2006, June). *Domestic Violence Analysis: Incidents Reported to Police in Anchorage, Alaska—Police Response and Incident Characteristics, Fourteen Year Study, 1989–2002.* Anchorage: Municipal Department of Health and Human Services.

National Academies of Sciences, Engineering, and Medicine (2016). *Modernizing Crime Statistics—Report 1: Defining and Classifying Crime.* Washington, DC: The National Academies Press.

National Academies of Sciences, Engineering, and Medicine (2017a). *Federal Statistics, Multiple Data Sources, and Privacy Protection: Next Steps.* Washington, DC: The National Academies Press.

National Academies of Sciences, Engineering, and Medicine (2017b). *Innovations in Federal Statistics: Combining Data Sources While Protecting Privacy.* Washington, DC: The National Academies Press.

National Academies of Sciences, Engineering, and Medicine (2017c). *Principles and Practices for a Federal Statistical Agency* (6th ed.). Washington, DC: The National Academies Press.

National Research Council (1976). *Surveying Crime.* Panel for the Evaluation of Crime Surveys. Bettye K. Eidson Penick (Ed.) and Maurice E. B. Owens III (Assoc. Ed.), Committee on National Statistics, Academy of Mathematical and Physical Sciences. Washington, DC: National Academy of Sciences.

National Research Council (2008). *Surveying Victims: Options for Conducting the National Crime Victimization Survey.* Panel to Review the Programs of the Bureau of Justice Statistics. Robert M. Groves and Daniel L. Cork (Eds.), Committee on National Statistics and Committee on Law and Justice, Division of Behavioral and Social Sciences and Education. Washington, DC: The National Academies Press.

National Research Council (2009). *Ensuring the Quality, Credibility, and Relevance of U.S. Justice Statistics.* Panel to Review the Programs of the Bureau of Justice Statistics. Robert M. Groves and Daniel L. Cork (Eds.), Committee on National Statistics and Committee on Law and Justice, Division of Behavioral and Social Sciences and Education. Washington, DC: The National Academies Press.

National Research Council (2014). *Estimating the Incidence of Rape and Sexual Assault.* Panel on Measuring Rape and Sexual Assault in Bureau of Justice Statistics Household Surveys. Candace Kruttschnitt, William D. Kalsbeek, and Carol C. House (Eds.), Committee on National Statistics, Division of Behavioral and Social Sciences and Education. Washington, DC: The National Academies Press.

New York Office of Justice Research and Performance (2016, April). *NYSIBR Data Capture Elements and Specifications.* Albany: New York State Division of Criminal Justice Services. Available: http://www.criminaljustice.ny.gov/crimnet/ojsa/crimereporting/ibr_refman/NYSIBR-Data-Capture-Elements-and-Specifications.pdf.

Norton, S. and D. Delay (2012, January). *Crime in New Hampshire 2011.* Concord: New Hampshire Center for Public Policy Studies.

Ohio Department of Public Safety (2017, March). *Ohio Incident Based Reporting System Data Collection and Submission Specifications.* Columbus: Ohio Department of Public Safety.

Oregon Uniform Crime Reporting Program (2017). *State of Oregon Report of Criminal Offenses and Arrests: 2016 Annual Report.* Salem: Oregon State Police.

Paoli, L. and V. A. Greenfield (2015). Starting from the end: A plea for focusing on the consequences of crime. *European Journal of Crime, Criminal Law and Criminal Justice 23*, 87–100.

Poggio, E., S. Kennedy, J. Chaiken, and K. Carlson (1985, May). *Blueprint for the Future of the Uniform Crime Reporting Program—Final Report of the UCR Study.* NCJ 098348. Washington, DC: U.S. Department of Justice, Bureau of Justice Statistics.

President's Commission on Law Enforcement and Administration of Justice (1967, February). *The Challenge of Crime in a Free Society.* Washington, DC: U.S. Government Printing Office. Commonly known as the Crime Commission report.

Public Administration Select Committee (2014, April 9). *Caught Red-Handed: Why We Can't Count on Police Recorded Crime Statistics.* Thirteenth Report of Session 2013–14; HC 760. Published by authority of the House of Commons. London: The Stationery Office Limited.

Reaves, B. A. (1993, October). *Using NIBRS Data to Analyze Violent Crime.* NCJ 144785. Washington, DC: U.S. Department of Justice.

Roberts, D. J. (1997, July). *Implementing the National Incident-Based Reporting System: A Project Status Report.* NCJ 165581. Washington, DC: U.S. Department of Justice, Bureau of Justice Statistics.

Sparrow, M. K. (2000). *License to Steal: How Fraud Bleeds America's Health Care System.* Boulder, CO: Westview Press.

Tennessee Bureau of Investigation (2013, October). *TIBRS Data Collection: An Instructional Manual for the Implementation of the Tennessee Incident Based Reporting System* (12th ed.). Nashville: Tennessee Bureau of Investigation.

Thompson, M. P., L. E. Saltzman, and D. Bibel (1999). Applying NIBRS data to the study of intimate partner violence: Massachusetts as a case study. *Journal of Quantitative Criminology 15*(2), 163–180.

Uniform Crime Reporting Program (2016, September). *Crime in Connecticut, 2015.* Middletown: Department of Emergency Services and Public Protection, Crimes Analysis Unit.

United Kingdom Government Statistical Service (2011, June). *National Statistician's Review of Crime Statistics: England and Wales.* London: United Kingdom Statistics Authority.

United Kingdom Statistics Authority (2010, May). *Overcoming Barriers to Trust in Crime Statistics: England and Wales.* Monitoring Report 5. London: United Kingdom Statistics Authority.

United Kingdom Statistics Authority (2011, April). *Assessment of compliance with the Code of Practice for Official Statistics: Crime Statistics in England and Wales (produced by the Home Office).* Assessment Report 102. London: United Kingdom Statistics Authority.

United Kingdom Statistics Authority (2014a, January). *Assessment of compliance with the Code of Practice for Official Statistics: Statistics on Crime in England and Wales (produced by the Office for National Statistics).* Assessment Report 268. London: United Kingdom Statistics Authority.

United Kingdom Statistics Authority (2014b, July). *Statistics on Recorded Crime in Scotland: Assessment of compliance with the Code of Practice for Official Statistics.* Produced by the Scottish Government. Assessment Report 288. London: United Kingdom Statistics Authority.

U.S. Department of Defense, Inspector General (2014, October 29). *Evaluation of the Defense Criminal Investigative Organizations' Defense Incident-Based Reporting System Reporting and Reporting Accuracy.* Report No. DODIG-2015-011. Washington, DC: U.S. Department of Defense.

U.S. Government Accountability Office (2011, June). *Organized Retail Crime: Private Sector and Law Enforcement Collaborate to Deter and Investigate Theft.* GAO-11-675. Washington, DC: U.S. Government Accountability Office.

U.S. Government Accountability Office (2017, September). *Costs of Crime: Experts Report Challenges Estimating Costs and Suggest Improvements to Better Inform Policy Decisions.* GAO-17-32. Washington, DC: U.S. Government Accountability Office.

Washington Association of Sheriffs and Police Chiefs (2017). *2016 Crime in Washington: Annual Report.* Lacey and Seattle, WA: Washington Association of Sheriffs and Police Chiefs and Washington State Criminal Justice Training Commission.

Wellford, C. (1982). Redesigning the Uniform Crime Reports. *American Journal of Police I*(II), 76–92.

Appendixes

– A –

Charge to the Panel on Modernizing the Nation's Crime Statistics

A panel of the National Research Council will assess and make recommendations for the development of a modern set of crime measures in the United States and the best means for obtaining them. For example, better information is needed on certain crime types such as against businesses or organizations and personal identity theft; also needed is greater ability to associate attributes such as firearms or drug involvement to crime types, and more complete adoption of electronic reporting, data capture, and system interoperability. The review will evaluate and make recommendations in the following areas:

- *Substantive*—Development of a framework for identifying the types of crimes to be considered in a modern crime classification, based on examining the strengths and limitations of various perspectives, such as: technical or legal definitions of crime types in criminal law or penal codes; definitions of "common unacceptable actions" in common law; and public health-type definitions that use affected persons (or victims) as the unit of analysis. The review will focus on full and accurate measurement of criminal victimization events and their attributes, considering types of crime (and their definitions), including the current scope of crime types covered by existing Federal Bureau of Investigation (FBI) and Bureau of Justice Statistics (BJS) data collections; gaps in knowledge of contemporary crime; development of international

crime classification frameworks that should be considered in increasing international comparability; and the optimal scope of crime statistics to serve the needs of the full array of data users and stakeholders—federal agencies, other law enforcement agencies, Congress, other actors in the justice system (such as the courts and corrections officials), researchers, and the general public;
- *Methodological*—Assessment of optimal methods to collect the data to complete the crime classification framework developed in the review, including assessment of the appropriateness of existing instruments and methods currently used by the FBI and BJS to collect crime information and the effectiveness and accuracy of their data processing techniques (including routines for imputation and estimation and the handling of missing data); the possible role of integrating data from non-FBI/BJS and nongovernmental sources (such as from credit card companies) into crime measurement; and capabilities for flexibly identifying and measuring new and emerging crime types going forward; and
- *Implementation*—How to maximize the use of locally collected and existing data as well as information technology assets while minding the voluntary nature of crime reporting in current systems, minimizing the effects of changes on state, local, and tribal law enforcement information management systems, meeting the needs of local law enforcement operations, and populating the national FBI and BJS data collections.

The review will also consider contextual information about crime produced by other statistical entities from different perspectives (such as contextual information on homicides, sexual assault, and stalking that may be derived from public health data collections), though the focus of the study is the taxonomy and measurement of crime and not the etiology of violence or deviant behavior generally. In addition, the review may consider cost-effectiveness and budgetary issues, such as priority uses for additional funding that may be obtained through budget initiatives or reallocation of resources among units of the U.S. Department of Justice.

The review will proceed in two stages. In phase one the panel will focus on the taxonomy and measurement of crime by hosting two workshop-style meetings, one with the research community and one with practitioners and policy makers, both addressing the scope and content of an ideal set of crime measures and indicators for the United States. Proceedings will be issued from both workshops, and the panel will produce an interim report that identifies lessons learned from the workshop input and directions for the panel's remaining work. In phase two, the panel will focus on the shape and structure of a modern set of crime measures—their sources, methods, tools, and processes—including what current sources of information could be used or modified to meet user needs, particularly for the higher priority measures. The panel will produce a final report at the end of the study that addresses ways to ensure that the nation has an integrated, complete, and contemporary set of indicators of the full range of crime (including the best means for disseminating data and findings) and document the joint role of FBI and BJS in producing those indicators.

– B –

Historical Themes in the Development of U.S. National Crime Statistics

We summarized the history of crime statistics in the United States in Report 1, albeit with a focus appropriate to the substantive focus of that volume: the meaning of crime and the range of offenses covered in crime statistics. But history affords valuable lessons, both aspirational and cautionary, in reinventing U.S. crime data collection, through emphasis on the establishment and implementation of existing systems. As context for discussion of the structure of a modern crime statistics program and its implementation, this appendix traces a number of important historical themes, among them:

- A transition from initial enthusiastic use of crime statistics as a tool for understanding the dynamics of crime to an inertial production-oriented mode that unduly limited the public concept of crime;
- Promising steps in more recent decades toward broader concepts of "crime" in national crime statistics and to measuring or estimating crime (rather than giving the appearance of completely enumerating it), including the emergence of the National Crime Victimization Survey (NCVS) and advent of the National Incident-Based Reporting System (NIBRS);
- Tendencies from the earliest days of national crime statistics collection to collect some additional information about crime beyond crude, total

offense count—as well as a tendency to greatly misjudge, also from those earliest days, the information resources available to local police departments and agencies; and
- Lessons learned from detailed operational reviews of crime statistics based on offenses reported to law enforcement in 1958 (when they were the only available national crime data) and 1984.

This historical review provides background that demonstrates how ingrained some concepts and challenges have become, and so is very useful to understanding the importance of making improvements to crime data collection such as those we propose in this study.

B.1 EARLY DAYS: CRIME STATISTICS AS TOOL FOR DISCOVERY

Crime is a weighty, serious, and painful topic, and rightly so. But, for aspirational purposes and to appreciate the importance of crime statistics as a basic social indicator, it is instructive to step back and consider the earliest days of nationally compiled crime statistics. One of the most striking things about leafing through the earliest publications of crime statistics is that crime data were presented and met with a sense of distinct enthusiasm—bordering on exuberance—at the new and major insights that the data could yield and the policies they could inform. The very first issue of the *Uniform Crime Reports* bulletin—the publication that would evolve into the annual *Crime in the United States*—sounded naturally cautious notes on the source and nature of the new data (Bureau of Investigation, 1930a:1–2):[1]

> All contributions of crime data [are] are purely voluntary, solicited solely for the purpose of making available statistical information concerning crime to those officially interested organizations and individuals who desire to receive and analyze the statistics. ... Every effort will be exerted by the Bureau of Investigation to secure from its contributors statistics which are not only accurate, but which will become more comprehensive as time goes on and the methods of compilation are more widely and thoroughly understood.

The early issues of the *Uniform Crime Reports* are replete with detail on the number of new police-department data contributors: marveling in the first issue at larger-than-anticipated initial participation by local departments ("the size of the registration area for crime reporting has more than doubled in the short period of eight months," with the number of cities returning one month or more of crime data growing "from 400 in January to a total of 895" in August;

[1]The bulletins would bear the full name *Uniform Crime Reports for the United States and Its Possessions* from 1930 through the first (of 2) issues for 1953, dropping "and Its Possessions" from the title thereafter. The first several issues were produced monthly, but that gave way to a quarterly cycle, then semiannual, and finally (in 1959) annual.

Bureau of Investigation, 1930a:1), celebrating the milestone of "every State of the Union" (then 48) "as well as the District of Columbia" being "represented" in the reports submitted in October 1930, by at least one local department (Bureau of Investigation, 1930b:1), and continuing as the number grew over time.

By the third bulletin in the series, the Bureau of Investigation analysts began assessing the content of the new data they had accumulated. "One of the purposes of collecting and publishing crime statistics is to show, as adequately as can be determined from the data available, the rise and decline in crime, or in certain selected crimes, over a given period of time," the bulletin noted. The analysis launched in the report was admittedly modest and crude: a look at 9 months of robbery and burglary data, disaggregated by population size of the reporting city. But it was insight that was newly possible—and so was taken with equal notes of caveat and enthusiasm. The bulletin cautioned that the two crime types were selected "because they are reported probably more widely and accurately than any other reportable offenses," and care was taken to further qualify the results (Bureau of Investigation, 1930b:4):

> The number of cities used [(695)] was determined by the number contributing returns on those offenses for the entire period [January–September 1930]. It does not purport therefore to include all cities in the country, nor even all that have at any time submitted a return. It is merely a miscellaneous group, drawn from cities of all sizes and in various sections of the country. It does include, however, approximately 80 percent of all cities of 100,000 population and over, 70 per cent of all cities of from 50,000 to 100,000 population. . . .

But, with those caveats, the results were striking for the day: putting into stark relief the sheer concentration of the volume of robbery and burglary offenses in the nation's largest cities (100,000 population or greater) and providing a first look at month-to-month fluctuation that would inform later study of seasonal effects of crime.

With the initial excitement over the possibilities of the new data came tensions over how much detail should be released and how uniformly concepts were being measured. In its first issue for 1932, the Bureau commented that "an effort has been made to present the statistics in a more desirable form"—dispensing with a city-by-city table of reported data due to "numerous requests for summarized data" (Bureau of Investigation, 1932:1); this level of detail was partially restored (for cities of population 100,000 or greater) by first quarter 1934, based on further feedback from the International Association of Chiefs of Police (IACP) advisory committee on the UCR. In 1935, puzzlement over the accuracy of homicide reporting led to the deployment of the first data collection to bear the name Supplementary Homicide Reports, this time "provid[ing] for the listing of detailed data on individual cases" of homicide on a short-term basis for selected cities (with population greater than 100,000). The supplemental collection revealed "that many cities were including, as actual offenses of murder, cases in which felons had been killed by a police officer in the line of duty," and so prompted formal change to protocols and specification of a standard definition for "justifiable homicide" (Federal Bureau of Investigation, 1939:70–71).[2] Early issues of the UCR bulletins are replete with concern about a potential point of ambiguity first expressed in the original UCR *Manual* (International Association of Chiefs of Police, 1929): the degree to which reported motor vehicle thefts were in fact true thefts of that specific type of property, rather than short-term instances of "joy-riding" and unauthorized use of a vehicle. Similarly, a 10th-anniversary history of the UCR program (Federal Bureau of Investigation, 1939:77) noted early and persistent concern over the consistency with which local police departments were heeding instructions that "all assault with a deadly weapon or with intent to kill should be listed as aggravated assaults;" that "simple assaults, assault and battery, fighting without weapons, and similar minor assaults" should *not* be so counted; and that ambiguous cases (such as "cases of assault with deadly weapons even though the assault was more or less the outgrowth of a fight") be handled consistently from month to month and year to year.

B.2 PUSH FOR MORE DETAIL, ALMOST FROM THE START

More to the point, enthusiasm over discoveries from shifts in the volume of crime types over time led to the push—within constraints of volunteer contribution of data from local police departments—for some finer categorization and detail. "In order to comply with suggestions received from police administrators[,] a supplementary report of known offenses" was distributed to large-city police departments (over 100,000 population) in January 1935; the first quarterly report for the year tallied "reports received from 19 cities" with a combined population of 3.7 million, as that small set of cities returned supplementary information that was "apparently complete and correct" (Bureau of Investigation, 1935:13). What the 1935 supplementary collection did was to divide most of the UCR Part I crimes into at least two finer subcategories, subdividing:

- rape as forcible or statutory;
- robbery by location (highway, commercial house, oil station, chainstore, residence, bank);

[2]Years later, a similar "special questionnaire" for select cities (over 25,000 population) was deployed in August 1960, to assess consistency of reporting of aggravated assault (Federal Bureau of Investigation, 1960:5). No detailed analysis is provided and no apparent change to definitions or concepts resulted, but the few summary statistics printed from the special study suggests initial, cursory study of the extent of later judicial follow-up on aggravated assault cases (e.g., whether charges were filed or convictions won).

- burglary by location (residence, all other) and time (during night, during day);
- larceny by value of article stolen (under $5, $5–50, over $50), by general type of property, or by mode (pocket-picking, purse-snatching, other); and
- auto theft as joyriding or all other.

To be clear, this was not collection of incident-level detail, but rather a separate accounting of offenses by the finer subcategories—and collection was initially limited to contributions by an unstructured sample of large police departments. Over the years, participation was requested from police departments in smaller cities, but still limited to only a relative handful of police departments, and some different categorizations added (e.g., basic weapon type used in homicide). A section tabulating these slightly-more-detailed data from "Supplementary Offense Reports" became a regular fixture in *Uniform Crime Reports* bulletins.

Eventually—and without substantial revision to the subcategorization since at least 1958—the "supplementary" information was folded into the regular UCR Summary data collection. In most cases, the subcategories first requested in 1935 became count subcategories on the basic "Return A" intended to be collected monthly from participating law enforcement agencies. Others developed into a permanent Supplement to Return A that remains part of the UCR Summary system, principally collecting summed estimates of the monetary value of property stolen (and property recovered by authorities) in the course of committing UCR Part I offenses.

B.3 RECURRING ASSUMPTION: THE DEPTH OF IN-HOUSE LOCAL LAW ENFORCEMENT DATA

The point of this appendix's historical discussion is to suggest themes in the development of national crime statistics. One such theme that recurs frequently is that planners and analysts tend to assume—often correctly, but sometimes radically overestimating—that local law enforcement agencies have ready access to highly detailed data and records internal to their own departments. In the early days of national crime statistics, this assumption was voiced by the FBI and lauded as a key feature of the national program because it ensured that local departments would easily be able to contribute. As the 10th-anniversary history of the program recounts (Federal Bureau of Investigation, 1939:46):

> It goes almost without saying that the practical basis of the uniform crime reporting system is one of its strongest features, inasmuch as all police executives should have constantly available for local administrative purposes up-to-date figures of the types to be included in the monthly and annual uniform crime reports. ... There are widespread evidences of a continuously growing and expanding professional spirit among the entire

> law enforcement personnel of the United States, particularly among police administrators, [in the compilation of] records and reports, and should contribute in future years toward the building of even higher standards of accuracy and uniformity.

Numerous UCR bulletins over the years anticipated the availability of richer local information in positing what it believed to be a likely use—and in some instances perhaps the most appropriate use—of the nationally compiled statistics: as a rough guide as to where to target analyses of those local police data to address the "local crime problem." One bulletin for 1934 (Division of Investigation, 1934a:9–10) took pains to emphasize that "the tabulation of crime rates for individual cities should not, in the opinion of the Division of Investigation, be used to discredit or to glorify individual police departments. They should serve as a source of information regarding crime conditions in individual cities which would otherwise be more or less unavailable to interested individuals." Instead, "it is believed that the following compilation may help to raise questions which will lead to further study of crime problems in individual cities which will indicate desirable remedial action." A second bulletin for 1934 went further, adding that "variations in the data for the several cities" actually heightened the importance of more detailed local study: "It is believed, therefore, that the proper function of the following compilation of crime data for cities with more than 100,000 inhabitants is to furnish the information which will serve as a stimulus to further study of local crime problems on the part of police administrators and interested civic organizations" (Division of Investigation, 1934b:9–10). Imparting more specifics, the final *UCR Bulletin* for 1953 (Federal Bureau of Investigation, 1954:72) anticipated that:

> A city police chief can compare his own figures for 1952 and 1953 and note the trend in his city, increases or decreases in the various crime classes. Then he can check his local trend data against the national figures given here in tables 25 and 26. This basic survey of trends gives direction to his additional studies of the local crime problem, such as: special types of robberies or burglaries; areas in his city with the greatest frequency of these crimes; time of day that they occur; possible rearrangement of patrol to combat the problem; and the like.

B.4 OSSIFICATION OF CONCEPTS, DIVERGENCE OF CRIME STATISTICS FROM THEIR USE, AND THE 1958 PROGRAM REVIEW

Just as UCR planners assumed that local law enforcement agencies might benefit from comparison of their local crime information with national numbers, they also recognized what the overriding interest in the nationally compiled data was likely to be. As Wellford (1982:81) put it in retrospect, "each agency is interested in its [own] performance. From the beginning the UCR[s] have been oriented toward the development of census procedures so that all agencies could be described in the report and so that agencies could review the data for jurisdictions and agencies with which they felt they could be compared." The operative phrase—and hint of the broader theme that took firm hold in the mid-20th century—is "census procedures." Over the decades, the concept of estimating crime and measuring it amidst uncertainty slipped away in favor of the appearance of completely enumerating it.

The earliest *Uniform Crime Reports* bulletins are remarkable for a lot of their language, as we have already described, but here it is useful to very briefly trace the evolution of two threads in those narratives. The first is a very real sense of the inherent limitations of the data flowing into the national level, which come out in occasionally blunt and candid statements. This began from the start, with the first issue of the *Uniform Crime Reports* bulletin and its concise description of the work-in-progress nature of national crime statistics, quoted above in opening Section B.1. By the following year (Bureau of Investigation, 1931b:1–2), the bulletin began urging that "caution should be exercised [in] arriving at any conclusions or deductions from the figures presented," for a variety of reasons. In addition to this being because "all cities of the country are not represented" in the tabulation of offenses known to the police, it was also because "wide divergencies and the total number of particular crimes for various cities of approximately the same population may in some cases not be indicative of a variance in the amount of crime in those cities." In turn, those differences "may be charged to inadequate record systems, or a lack of understanding of the [UCR] classification." The very next issue of the bulletin (Bureau of Investigation, 1931a:1) boiled things down further still, into a terse disclaimer—actually, more of a disavowal—that was repeated in several subsequent issues before being moderated somewhat:

> In publishing the reports sent in by chiefs of police in different cities, the Department of Justice does not vouch for their accuracy. They are given out as current information which possibly may throw some light on problems of crime and criminal law enforcement.

At the same time, the *Uniform Crime Reports* Bulletins began an escalating, internal rhetorical battle of sorts against that primary projected use of the compiled crime data: comparison between jurisdictions and over time. When raised as a concern for the first time in 1935, the discourse was remarkably polite (Federal Bureau of Investigation, 1935a:13):

> The Federal Bureau of Investigation desires to point out that comparisons between the figures of two or more cities should be made with caution. Just as peculiar local conditions may have a tremendous bearing upon the mortality rate, so similarly peculiar local conditions may cause the crime rates to be above or below average. In other words, in making comparisons

> with the average figures or the figures of a single community, consideration should always be given to local conditions which may have a large amount of influence with regard to causing the amount of crime to be unduly large or unusually small.

UCR writers and editors clearly grappled with how they wanted to present the point, with the very next bulletin simplifying the advisory to the single sentence (Federal Bureau of Investigation, 1935b:12): "Comparisons between the figures of two or more individual cities should be made with caution because there may be present any number of peculiar local conditions which may cause the crime rates to be above or below average." By 1940, though, the sentiment seems to have switched to the advisory not being specific enough about relevant "peculiar local conditions." In that year, the disclaimer expanded (Federal Bureau of Investigation, 1940a:11):

> A great deal of caution should be exercised in comparing crime data for individual cities, because differences in the figures may be due to a variety of factors. The amount of crime committed in the community is not solely chargeable to the police but is rather a charge against the entire community. The following is a list of some of the factors which might affect the amount of crime in a community:
>
> - The composition of the population with reference particularly to age, sex, and race.
> - The economic status and activities of the population.
> - Climate.
> - Educational, recreational, and religious facilities.
> - The number of police employees per unit of population.
> - The standards governing appointments to the police force.
> - The policies of the prosecuting officials and the courts.
> - The attitude of the public toward law-enforcement agencies.
> - The degree of efficiency of the local law-enforcement agency.
>
> Comparisons between the crime rates of individual cities should not be made without giving consideration to the above-mentioned factors. It is more important to determine whether the figures for a given community show increases or decreases in the amount of crime committed than to ascertain whether the figures are above or below those of some other community.

And, by 1950, the bulletin took pains to present, as a paragraph of underlined text, this caution (Federal Bureau of Investigation, 1950:13):

> Caution should be exercised in comparing crime data for individual cities because the differences in the figures may be due to a variety of factors. Such comparisons are not desirable even though the figures for individual communities are converted into terms of the number of offenses per 100,000 inhabitants.

As a coda (albeit not underlined), the warning returned to the same "middleman" argument that had been advanced as early as 1931: "In publishing these figures the FBI acts as a service agency. The figures published are those submitted by the contributing agencies."

Nearing the 30th anniversary of the UCR Program's establishment, the FBI commissioned a three-member expert Consultant Committee on Uniform Crime Reporting in 1957 to review the conduct of the UCR Program;[3] its results were published in a special issue of the UCR bulletin in 1958 (Federal Bureau of Investigation, 1958), and we summarize their recommendations in Box B.1. Though its discussions were limited principally to a set of specific issues raised by the FBI, the committee's review was very thorough and touched on some of the themes raised heretofore in this appendix and report. In particular, the committee noted "a regrettable amount of misinterpretation by some" upon every release of UCR numbers, "[causing] unnecessary concern on the part of local governmental agencies, private organizations, and the public, frequently about imaginary problems" (Federal Bureau of Investigation, 1958:11). The committee expressed particular worry about the index/Part I crimes being misconstrued as reflecting "all crime" and suggested that part of the problem was the UCR Program's delineation between Part I and Part II offenses. The argument was, essentially, that distinction between the "serious" Part I crimes and the set of Part II offenses implies that Part II is nonserious. Instead, should a distinction have to be made, the committee suggested that the line be drawn between crimes that manifest or appear first through report to the authorities and those that manifest or appear first through the making of an arrest. As to how to collect the data, the committee was not worried about the core "offenses known to the police" (Part I offenses) because, "assuming that the trend toward increased coverage continues, more or less complete reporting will be a reality in a very few years" (Federal Bureau of Investigation, 1958:39). But those offenses that are first seen through arrest (typically, the Part II offenses but potentially including others) suffered from irregularity in data collection. The committee suggested that either concerted effort be taken to bring participant reporting of Part II offenses up to the Part I norms or that estimating Part II offenses through a sampling procedure be considered instead. In publishing the committee's report as a special edition of the *Uniform Crime Reports* bulletin, the FBI noted general agreement with the principles embodied in its recommendations—most notably the call to cut from semiannual reporting to a single annual volume.

But the actions taken beginning in 1959 belied much of the guidance of the expert committee and exacerbated some of the tensions and developments

[3]The committee was chaired by sociologist Peter Lejins, University of Maryland, and its members were Charlton Chute, Institute of Public Administration, New York City, and Chief Stanley Schrotel of the Cincinnati Police Department.

Box B.1 Recommendations of Commissioned Expert Panel Review of Uniform Crime Reporting Program, 1958

1. The most enduring result of the 1958 committee's work was the FBI's subsequent rebranding of the annual crime statistics report as *Crime in the United States*, coupled with an emphasis on the Part I "index crimes." But what the committee actually recommended was considerably more nuanced:
   - The committee argued for a *deemphasis of the FBI's distinction between "Part I" and "Part II" labels*. "Confusion would result" if those labels were retained, and "new terms applied to a new classification would promote clarity." (However, the committee separately recommended that the FBI *retain the Part I and II labels "at least for the time being" if deemed useful for internal communications with participating police departments*.)
   - The committee suggested a twofold primary "objective of publishing statistics on criminal offenses," *comprising both "a meaningful index of crime for the United States (presently accomplished by reporting of [Part I] offenses)" and "the total volume of criminal offenses committed in the United States, by categories, perhaps entitled General Crime Statistics for the United States."* "It is understood that both of the statistical series are to serve as the basis for the computation of appropriate rates and trends and for making comparisons."
   - Concerning the "crime index," the committee did not suggest any major revision of the designated Part I offenses, but did suggest that *"the tabulation should be entitled 'index of crime' and the composition of the index be changed to exclude manslaughter by negligence and larceny below $50."* Two subsequent recommendations reiterated that these specific offenses should be excluded from the index, but that *manslaughter by negligence and larceny below $50 should "continue to be included in the United States crime statistics"*—just not as part of the index.
   - In lieu of "Part I" and "Part II," the committee actually argued for a *"classification" of the tabulations of crime data based on how the underlying offenses came to be known.* Specifically, the tabulations "should be divided into those offenses which are reported as they become known to the police and those which are reported as they become known through the arrest of the offender." (A separate recommendation urged that this *delineation of known-by-report offenses and known-by-arrest offenses be "further studied by the personnel of the [FBI] in consultation with experienced representatives of the police departments in order to determine the best manner of handling specific offenses."*)
   - Two further recommendations further emphasized the importance of the so-called Part II offenses. The committee argued that *the aim of a "total crime picture for the United States" must involve a "tabulation including all offenses, perhaps with the exception of some minor ones"*, explicitly reaffirming that this meant the offenses covered under Part II as well as those in Part I. Moreover, the committee emphasized the need for *improved measures of "Part II" offenses*—including "intensif[ying]" data collection "in order to achieve more complex coverage for the urban communities, and a program of collecting this information for the rural areas should be initiated" so that all might be analyzed with data disaggregated by such factors as age, sex, and race.
2. While commending the FBI for having achieved "remarkable coverage" of Part I offenses known to law enforcement, the committee suggested that the Bureau should follow "1 of 2 courses" for all tabulations not based on such high levels of coverage: *"either a definite attempt should be made to build up the reporting to achieve more or less complete coverage" or "a sampling technique should be resorted to"* and estimates generated based on the sample.

Box B.1 (continued)

3. Arguing that the Uniform Crime Reports (UCRs) "have outgrown being of use primarily to law enforcement" and rather "have become a basic source of information on crime for the country as a whole," the 1958 committee urged that the program's reports be restructured to *emphasize the broader array of users (and uses) of crime data.* Specifically, the statistics should be presented "in such a way that the information will be meaningful to the general public and interested agencies and organizations even outside the immediate field of law enforcement."
4. Recognizing that meeting this broader focus might require additional time for tabulation and analysis, the committee urged that the FBI *reduce publication frequency to a single, annual report* rather than two reports a year. (That said, "issuance of intermediate publications of limited scope, which the FBI might find appropriate for release, is of course not precluded by this recommendation but is actually encouraged.")
5. Along similar lines, the committee urged that "*the objectives for collecting and presenting the statistical information on criminal offenses should be clearly formulated and adhered to at all times,*" including printing a "brief restatement" of said objectives in each subsequent UCR data report "to forestall misinterpretation of the data."
6. While emphasizing the need for the UCR Program to consider non-law enforcement users, the committee also suggested that *a special section in the crime statistics cover "the group of offenses which at a given time are especially important to the police in its work,* so as to increase the effectiveness of the police in combating the particular offenses by focusing attention on them and providing additional information on their frequency and distribution and their success in suppressing them" Tentatively, the committee suggested that manslaughter by negligence be one of these special offenses; for another, the committee suggested that the FBI examine "the traffic-fatality data collected, tabulated, and published by the National Safety Council," perhaps establishing a continued measure of traffic fatalities "provided the source is competent and reliable."
7. "In making changes to the UCR's, *the utmost care should be taken not to destroy the continuity of the statistical series*. Whenever feasible, both versions, i.e., the old and the new, should be given, or the data should be presented in such a way that the reader himself can make the calculations necessary to utilize the data earlier collected and published."
8. In four related recommendations, the committee expanded upon their perceived need for *analysis of crime data to differentiate between urban and rural crime*. The committee suggested a *basic classification of geographic areas into metropolitan areas, urban communities with more than 2,500 inhabitants, and rural communities with population less than 2,500* as defined by the decennial census. In the reports, *"Rural Versus Urban Criminality" should be tabulated separately*–and, "in view of the remarkable coverage for offenses known to the police," the *contrast between urban and rural levels of "index crime" should be "Presented more prominently and more accessibly."* Finally, and more technically, the committee urged the FBI to *use the Census Bureau's (intercensal) population estimates to define the urban and rural areas* rather than carry over the most recent decennial census data for a full 10 years.
9. The committee urged *examination of longer-term trends in offenses through "comparisons with more than only the previous year."*
10. *"A permanent technical consultant committee should be established to be available to the Bureau staff for consultation whenever needed."*

SOURCE: Adapted from recommendations section in Federal Bureau of Investigation (1958).

recounted earlier. Concern having been expressed about Part I-heavy, UCR data being unduly portrayed as a full accounting of crime, two things in the opening pages of the very next issue of UCR data in 1959 are interesting to note: *Crime in the United States* emblazoned on the cover of the volume in bold capital letters[4] and a new summary section conveying highlights of "crime volume" and "crime totals" with little to no condition being put on what was meant by "crime" in those assessments. Granted, the new summary began with a preamble that properly credited the voluntary nature of crime data contribution—but did not really describe the limitations of the data and, worse, characterized the actual UCR tabulations as almost superfluous detail:

> You have here a wealth of information on crime in the US, due to the cooperation of police and sheriffs who voluntarily send crime figures to the FBI.
>
> This summary is for the reader interested in the general crime picture. Technical data, of interest primarily to police, social scientists, and other students, are presented in the following sections.

Reports of the era continued the unusual practice of touting the top-line summary totals while, again, signaling a distinctly hands-off approach to the processes by which data were submitted. To wit, part of the boilerplate disclaimer that ran in *Crime in the United States* volumes beginning in 1961 reads (Federal Bureau of Investigation, 1961:27–28):

> It is clear, of course, that regardless of the extent of the statistical verification processes used by the FBI, the accuracy of the data assembled under this program depends upon the degree of sincere effort exerted by each contributor to meet the necessary standards of reporting, and, for this reason, the FBI is not in a position to vouch for the validity of the reports received.

What followed for the next several decades was fairly predictable. Wellford (1982:78) lamented that "UCR data have been used by some as if the data were a valid social indicator of the total incidence of crime," even though "the UCR was not created to be such a social indicator," generally. But a major reason for this is not hard to intuit—branded as *Crime in the United States*, the UCR Summary data came to *define* "crime in the United States," to reduce the national conversation on crime mainly to the reporting of short-term changes in a few, broad categories of crime rates, and to lend themselves to less-than-careful comparisons across jurisdictions because they were all that were available. Clear statements of the uncertainty associated with UCR figures have never really been attempted, and what few hints about individual agencies' participation (in part or in full) over the reporting years were largely obscured over time. Even today, the UCR data publications themselves do

[4] Internal pages still bore the usual Bulletin nomenclature, but the report itself was eventually renamed to fit the new cover name.

little to illuminate the error properties of the police-report data, with some clues peppered into footnotes (in the days of paper publication) or auxiliary documents (the separate "data declarations" accompanying the more recent electronic releases of *Crime in the United States*).

B.5 EMERGENCE OF THE NATIONAL CRIME VICTIMIZATION SURVEY

The *Crime in the United States*-as-"crime in the United States" effect is pervasive and still continues in some form, though a very powerful corrective to the UCR as the only available measure of crime began its evolution in the late 1960s. A prototype survey (Ennis, 1967) fielded in support of the work of the President's Commission on Law Enforcement and Administration of Justice (1967) demonstrated the potential of crime measurement through survey interviews, and led to the implementation and development of the National Crime Surveys (NCS) in the 1970s.[5] In its design—largely to ensure direct comparability with the UCR's police-report data—the new NCS deliberately adhered to some of the UCR's definitions and concepts. But it was revolutionary in providing a means to quantify the vast amount of crime *not* reported to the police and shed light on the so-called "dark figure of crime" that is, by definition, excluded in the *Crime in the United States* police-report tallies.

Sponsored by the Bureau of Justice Statistics (BJS) and using the U.S. Census Bureau as the data collection agent, the survey was redesignated as the National Crime Victimization Survey (NCVS) after a comprehensive redesign in the early 1990s. The NCVS has expanded its reach over the years through the implementation of topic supplements and revision of core content (for instance, querying respondents about experiences with identity theft or with their interactions with the police or other actors in the criminal justice system).

Among the reasons why we do not devote a great deal of page-space in this report to the NCVS and its conduct is that it has been the principal focus of several previous National Academies studies. The panel study that produced National Research Council (1976) served as a comprehensive early-phase review of the surveys; our predecessor panel to review the full portfolio of BJS programs dedicated a first report volume exclusively to the NCVS (National Research Council, 2008) and considered the NCVS's fit into that portfolio in its final report (National Research Council, 2009); and a study of the measurement of rape and sexual assault in federal surveys (National Research Council, 2014) considered the implicit limitations of an "omnibus" survey vehicle like

[5]It is "Surveys," plural, because the survey program initially included a separate sample of businesses as well as essentially stand-alone surveys for some individual surveys; these did not long survive initial review, and the program consolidated behind the single National Crime "Survey" of households thereafter.

the NCVS in eliciting information about personally sensitive victimization experiences. More generally, going beyond the National Academies' projects, the reason why we spend relatively little time talking about the NCVS in modernizing national crime statistics is that "modernization" is effectively the continual norm for the survey. Though it has endured seasons of budget-induced cuts in sample size, the NCVS is almost constantly in a state of evaluation and development related at improving the survey's efficiency. This includes active research and development to find ways to overcome the chief liability of the NCVS, which is that most respondents reached by the survey at an given time will not have experienced crime in the past 6 months, much less experienced highly specific offense types. Hence, because specific instances of crime victimization are (statistically) a relatively rare event in the whole population, the NCVS is necessarily limited in the degree to which it can provide estimates of specific offense types or victimization measures among particular demographic or geographic subgroups. Nevertheless, BJS continues research work on generating some regular subnational estimates from the NCVS.

B.6 THE *BLUEPRINT* AND THE DEVELOPMENT OF THE NATIONAL INCIDENT-BASED REPORTING SYSTEM

In the early 1980s, BJS sponsored a task force study to reimagine the collection of police-report data. The centerpiece of that task force's work was a conference on the future of the UCR, held in early 1984, and culminated in a final report entitled the *Blueprint for the Future of the Uniform Crime Reporting System* (Poggio et al., 1985). The task force used the terminology "unit-record reporting" to describe its primary vision for next-generation UCR data collection, though the term that ultimately took hold would be "incident-based reporting." The task force's recommendations are shown in Box B.2 and suggest—like the 1958 committee before it—a thoroughgoing review of operations but one that stopped short of considering crime data collection more generally. It did not challenge the basic content of the UCR Summary system but it did add considerably to the number of covered offenses and the level of detail requested about each incident for several offense types.

In the wake of the *Blueprint*'s release, pilot work began with several law enforcement agencies in South Carolina in 1987, and new concepts for the proposed National Incident-Based Reporting System (NIBRS) were announced at a national UCR conference in March 1988. But the *Blueprint* task force shares another distinction with the 1958 UCR committee that preceded it: Its central recommendations were followed only in part, and not exactly in the intended direction. As outlined in Box B.2, the task force had incident-based reporting as an aspirational goal for all law enforcement agencies, but took care

Box B.2 Recommendations of the Uniform Crime Reporting Study Task Force, 1984

3.1 Convert the UCR system to a two-level reporting system under which most agencies report basic offense and arrest information similar to that currently reported, while a comparatively small sample of agencies report much more extensive information.
4.1 Convert the entire UCR system to unit-record reporting in which local law enforcement agencies submit reports on each individual criminal incident.
4.2 Convert the entire UCR system to unit-record reporting in which local law enforcement agencies submit reports on each individual arrest.
5.1 [Level I reporting agencies only] Retain data collection for Part I offenses only, but eliminate negligent manslaughter altogether and broaden the rape category to include all forcible sexual offenses in Part I.
5.2 [Levels I and II] Distinguish attempted from completed offenses.
5.3 [Levels I and II] Report other distinct offenses occurring within a criminal incident, in addition to the most serious offense as determined by the Hierarchy Rule; retain the Hierarchy Rule to determine the most serious offense for each victim within a criminal incident.
5.4 [Levels I and II] Redefine aggravated assault more explicitly in terms of the use of weapons and the extent of injury.
5.5 Collect additional information about homicides.
5.6 [Levels I and II] Distinguish among crimes against businesses, crimes against individuals or households, and crimes against other entities.
5.7 [Levels I and II] Distinguish crimes against residents of a jurisdiction from crimes against nonresidents, in order to be able to adjust for large influxes of nonresidents either as daytime business populations or as tourists.
5.8 [Levels I and II] Collect value of property stolen by dollar value and provide for the value to be indicated as missing for cases in which it is not known.
5.9 [Levels I and II] Record related incident numbers on each arrest report and submit reports on exceptional clearances, in order to increase the accuracy of clearance data.
6.1 Seek participation in the Level II component from all agencies serving populations in excess of 100,000 and a sample of at least 300 smaller agencies.
6.2 [Level II] Collect Part II, as well as Part I, offense data and use more detailed offense-type categories than the current categories.
6.3 [Level II] Collect detailed incident data describing the nature of the criminal incident, including victim and offender characteristics, victim-offender relationship, use of force, nature and extent of injury, and type of location.
6.4 [Level II] Collect data periodically describing the characteristics and policies of each reporting law enforcement agency and assemble these data together with demographic, socioeconomic, and physical characteristics of the jurisdiction, which should be obtained from other sources such as the U.S. Census Bureau.
6.5 Design the National Program to allow for a variety of levels of state program participation in Level II.
7.1 [Levels I and II] Institute routine, ongoing audits of samples of participating UCR agencies in order to establish the extent of error in the system on a continuing basis.
7.2 [Levels I and II] Develop a code of professional standards for reporting systems.
7.3 [Levels I and II] Develop improved feedback to agencies through self-administered proficiency tests, periodic reports of common audit errors, and regular reports to individual agencies on the extent of edit discrepancies in their UCR submissions.
7.4 [Levels I and II] Strengthen state UCR program quality assurance, including expansion of local agency audits conducted by state programs.

Box B.2 (continued)

8.1 Develop the UCR, the [NCVS], and [Offender-Based Transaction Statistics [(OBTS)] systems as independent programs providing complementary criminal justice statistics for multiple purposes. The strengths of each of these data systems should be continued and enhanced, rather than compromised to achieve face comparability.
8.2 Structure the UCR and [NCVS] data so as to permit reconciliation of the two.
8.3 Develop data structures and associated audit procedures with an eye toward eventual analytic integration of the estimation of crime rates and trends from UCR and [NCVS] data.
8.4 Design the UCR system to allow linkage of police records to the prosecution and court records collected by OBTS systems.
9.1 Create six publication series, including:
- an annual report that is basically factual but more textual and interpretive than the current report;
- quarterly releases of crime counts and trends;
- annual compilations of statistics for local jurisdictions, similar to those currently in *Crime in the [United States]*;
- a series of computer-generated special reports to individual agencies or groups of similar agencies;
- a series of occasional publications analyzing special issues about crime, primarily directed at researchers; and
- a series to provide for publication of methodological details and technical documentation.

9.2 Issue UCR reports at least once a year jointly with a corresponding report from the National Crime [Victimization] Survey, and occasionally issue joint publications.
9.3 Provide a continuing analysis capability for reconciliation of UCR and [NCVS] data, evaluating seriousness scoring, and preparing periodic publications, special studies, and technical documentation.
9.4 Support continued and enhanced user services, including a user data base with files linked over time, the capacity to draw samples of offenses for analysis either by the UCR staff or by outside researchers, in response to public queries.

SOURCE: Excerpted from Poggio et al. (1985).

to define two levels of participation. First-phase implementation of incident-based collection would be targeted at a sample of law enforcement agencies (the nation's largest and hence, arguably, its best resourced police departments, plus a representative sample of smaller departments). However, in implementing the new system, the FBI opted to open the database and its new format to purely voluntary contribution from the full set of usual UCR participants.

The FBI's first revision of its *UCR Handbook* after the establishment of NIBRS—the "NIBRS Edition" of 1992—did a commendable job of explaining the new program. Indeed, 25 years later, it is difficult to craft a better capsule summary of the potential benefits of NIBRS than that advanced in the handbook and quoted in Box B.3. Yet that statement also includes a strong instance of the recurring assumption noted earlier in Section B.3: greatly overestimating either the depth of data in local agencies' internal databases or their willingness/ability to share those data. On how NIBRS would actually come into existence, the manual observed that "the implementation of NIBRS will be at a pace commensurate with the resources, abilities, and limitations of the contributing law enforcement agencies" (Federal Bureau of Investigation, 1992:1). Like other UCR/NIBRS-related documents from the early 1990s, the manual sounded very optimistic notes on NIBRS take-up. One page after the comment on implementation pace, the UCR handbook offered reassurance that "*Crime in the United States* will continue to be published in its current format until conversion from the traditional summary UCR system to NIBRS is complete in most of the country"—implying that the wait would not be excessively long. Meanwhile, BJS's first bulletin analyzing NIBRS data, *Using NIBRS Data to Analyze Violent Crime* (Reaves, 1993, using 1991-vintage data), included a foreword projecting that "an estimated 40% of the Nation will report to NIBRS by the end of 1994." Both the FBI handbook and the BJS bulletin seemed to project quick take-up rates based on greater-than-expected submissions (and promised submissions) of test data from several states, as well as full-fledged early adopters such as South Carolina, Alabama, and North Dakota—as well as the critical assumption that NIBRS was simply going to be able to take advantage of crime data already residing in local law enforcement records management systems. To wit, the UCR handbook commented (Federal Bureau of Investigation, 1992:1):

> NIBRS data are designed to be generated as a byproduct of local, state, and Federal automated records systems. Thus, an agency can build its own system to suit its individual needs, including all the information required for administrative and operational purposes. Only the data required by NIBRS are then reported to the national UCR Program.

In practice, though, take-up rates have been much slower. As of 2016, NIBRS had yet to achieve the projection of 40 percent participation, whether measured as a percent of population coverage or by proportion of UCR-reporting agencies submitting NIBRS data, that was originally forecast for 1994. In 1996, the FBI received NIBRS-format data representing 1,550 departments—under 10 percent of the more than 16,000 law enforcement agencies then contributing to UCR programs. For a system that the *Blueprint* authors suggested should focus first and heavily on the largest police departments, those 1,550 departments included only one with a service population of more than 500,000: Austin, Texas (Roberts, 1997). In subsequent years, BJS's grant programs offered assistance to some additional state agencies to plan and implement NIBRS-format collection at the state and local levels. But a major consequence of the limited NIBRS take-up was that the nationally compiled NIBRS data could not be said to be "national" at all: that is, they could not be deemed statistically representative of the nation as a whole or, ultimately,

Box B.3 "Benefits of Participation" in the National Incident-Based Reporting System, 1992

An indispensable tool in the war against crime is the ability to identify with precision when and where crime takes place, what form it takes, and the characteristics of its victims and perpetrators. Armed with such information, law enforcement can better make its case to acquire the resources it needs to fight crime; and after obtaining those resources, use them in the most efficient and effective manner. NIBRS provides law enforcement with that tool because it is capable of producing more detailed, accurate, and meaningful data than produced by the traditional summary UCR Program.

Many individual law enforcement agencies have very sophisticated records systems capable of producing the full range of statistics on their own activities. NIBRS allows common denominator links among agencies. It will provide law enforcement agencies with extensive, specific crime information concerning similar jurisdictions, allowing the identification of common crime problems or trends. Agencies can then work together to develop possible solutions or proactive strategies for addressing the issues.

Legislators, municipal planners/administrators, academicians, penologists, sociologists, and the general public will be better able to assess the Nation's crime problem using the extensive data supplied by the law enforcement community. Law enforcement is a public service and as such requires a full accounting from the police commissioner, police chief, sheriff, or director as to the administration of the agency and the status of public safety within the jurisdiction. Full participation in NIBRS will provide statistics to enable a law enforcement agency to fulfill this responsibility. NIBRS has, in fact, the capability of furnishing information on nearly every major criminal justice issue facing law enforcement today, including terrorism, white-collar crime, weapons offenses, missing children where criminality is involved, drug/narcotics offenses, drug involvement in all offenses, hate crimes, spouse abuse, abuse of the elderly, child abuse, domestic violence, juvenile crime/gangs, parental kidnaping, organized crime, pornography/child pornography, driving under the influence, and alcohol-related offenses. The data will be available from all levels of law enforcement—Federal, state, and local—aggregated at the level and in the manner which best meets the informational needs of the data user.

SOURCE: Excerpted from Federal Bureau of Investigation (1992:3).

to major geographic or demographic subpopulations. Though BJS and some researchers[6] performed limited analysis of NIBRS-compiled data—hinting at the analytic possibilities—NIBRS has lacked the ability to "sell itself" based on the utility of its data, and it took until 2013 for the FBI to issue a full report and release (however incomplete) of NIBRS data.

We will return in detail to the barriers to NIBRS adoption in Section D.1, but one contemporaneous summary warrants mention in this historical narrative. Commissioned by BJS, SEARCH (the National Consortium for Justice Information and Statistics) convened ad hoc surveys and focus groups in the late 1990s to try to document the obstacles to NIBRS adoption. The

[6]See, e.g., Reaves (1993); Thompson et al. (1999); Addington (2007); Chilton and Regoeczi (2007); Haas et al. (2007).

cost of converting to NIBRS-format reporting was the most cited barrier; the focus groups also spotlighted concerns about artificial "spikes" in local crime that would be created simply by the switch from reporting multiple offenses rather than a single offense (under the UCR Hierarchy Rule). Law enforcement executives also expressed concern that completing incident-based reports might be too large a time drain for officers—a concern related to a lack of clarity concerning the direct benefits of national data compilation to reporting agencies. Also tellingly, the focus groups suggested more structural barriers: respondents noted concern about the arduousness of meeting FBI specifications for certification to submit NIBRS-type data, and several expressed a basic worry that federal officials were not entirely committed to developing NIBRS (Roberts, 1997; Maxfield, 1999).

B.7 THE STATUS QUO: "NEW" CRIMES IN "OLD" STRUCTURES

On the UCR/police-report side of national crime statistics, a firm axiom developed over time, being restated several times by the FBI at several meetings of our panel: The only way to alter the content of the UCR Program is either through formal recommendation of the FBI's Criminal Justice Information Services (CJIS) Advisory Policy Board (APB) or by enactment of federal law. Both of these channels for effecting change have strong or compelling arguments behind them, not least of which is that the multitiered APB process—vetting proposals and policies through an elaborate set of regional and topic-specific working groups (including a dedicated UCR subcommittee)—ensures that the perspectives of numerous law enforcement stakeholders, at least, are reflected in board recommendations.

What warrants mention as a historical theme, and lesson for moving forward, is that both of the principal channels for change in the nation's current crime statistics have major and perhaps unintended consequences for the perceived integrity of those statistics. More specifically, the APB process is so cautious as to appear sluggish or unresponsive in answering questions of high national interest, and the federal legislation route forces "new" (and very different) offense types into "familiar" UCR (and, to a lesser extent, NCVS) templates at the expense of accuracy.

The signature example on the APB side is the decades-long effort to modernize the UCR's definition of rape, which had been set in 1929 as "the carnal knowledge of a female, forcibly and against her will" (International Association of Chiefs of Police, 1929:195). Specific proposals to make the definition gender-neutral and to remove the "forcible" concept date back to, at least, the *Blueprint* review of the UCR Program in 1985 (see Box B.2). Yet even simple wording changes in definition can be costly to implement at the local level, and difficulty of implementation (including reconciliation with state criminal code and recording practices) undoubtedly slowed the process. Ultimately, it would take until 2011 for the APB to recommend and for the FBI and Justice Department to approve a revised definition. But other examples are also telling: for instance, it took until 2015—very well into the computer age—for hacking/computer invasion to be added as a NIBRS offense category. This is perhaps not surprising given an observation repeated by several of the users/stakeholders the panel engaged in its workshops: The content of the Supplement to Return A (see Box C.2) has not been systematically reviewed since at least 1958. Accordingly, the limited offense and property loss estimates available through the UCR Summary have long been sorely out of date because they do not include at least one very major target of modern larceny and theft: consumer electronics that do not neatly fit the category heading of "televisions, radios, stereos, etc."

The simplest way to appreciate the way in which legislative mandates have been shoehorned into existing UCR templates and structures is to review the list of forms and data collections in the UCR in Box 1.1 (Chapter 1). Four of the most recent and most important legally mandated UCR topics are almost literally tacked on, as separate forms, to the UCR Program. The addition of arson (1978; 92 Stat. 933), bias-motivated offenses/hate crimes (1990; 104 Stat. 140), cargo theft (2006; 120 Stat. 240), and human trafficking (2008; 122 Stat. 5083) each occasioned the development of a parallel data-reporting form as an add-on to existing UCR Summary collection. New codes for these offenses and for bias motivation could more readily be added to NIBRS though not—as in the rape definition revision—without difficulty. Typically, the new forms were also deployed to UCR respondents without much (if any) training in how to recognize relevant incidents and record them properly. The NCVS has been subject to similar change through legislation, as described in Section 2.2.1 of Report 1 and more completely by National Research Council (2008)—the most recent and arguably the most curious one being the provision etched into fiscal year 2015 appropriations law requiring the NCVS to produce estimates of honor violence.

The coda to this historical theme is that the current channels by which national crime statistics can be changed are predisposed to try to fit "new" and difficult concepts into "old" structures, rather than begin with the question of what is the most appropriate and accurate method by which statistical information on the topic could be derived.

B.8 NATIONAL CRIME STATISTICS EXCHANGE, SUNSETTING THE SUMMARY COUNTS, AND THE PATH AHEAD

An important but as-yet unspoken theme from the history of crime statistics is that the UCR's police-report statistics are of great and natural importance to BJS's central mission, even though the UCR is administered by the FBI. So it was neither accident nor typographical error to say that BJS sponsored the *Blueprint* review of the future of UCR in the early 1980s, that it commissioned SEARCH's investigation of the barriers to NIBRS adoption, and that it tried to encourage NIBRS adoption through grant monies in the 1990s and 2000s. This is due in large part to BJS's administrative home in the Office of Justice Programs (OJP) in the U.S. Department of Justice, which in turn inherited the functions of the former Law Enforcement Assistance Administration (LEAA). Put simply, BJS and OJP have the capacity and mission to provide grants and technical assistance that the FBI as a law enforcement agency generally does not.

Accordingly, one of the most recent, positive developments in moving toward modern national crime statistics is the National Crime Statistics Exchange (NCS-X), announced by BJS in 2012 with funding from the FBI. NCS-X leverages BJS's grantmaking authority to directly fund local and state agencies to implement NIBRS-compliant information management systems, with the first targeted efforts applied to a carefully designed sample of jurisdictions. When complete, and those new NIBRS participants' data are combined with data from existing NIBRS reporters, the resulting dataset will finally be nationally representative. In this way, analysis of national crime dynamics will finally be able to benefit from standardized cross-jurisdictional data to inform solutions—and, in principle, convincing other local agencies of the virtues of NIBRS participation should be easier. This is, in some sense, a return to the original *Blueprint* vision of seeding detailed data collection in an initially small sample of jurisdictions, to permit the data system to function fully as other law enforcement agencies switch to incident-based reporting over time.

Early NCS-X activities included interviews with state UCR programs, a survey of all 400 sampled agencies, and readiness assessments and coordination with both state programs and sampled agencies. Importantly, the early work also included outreach to pivotal groups including the Association for State Uniform Crime Reporting Programs (ASUCRP), the IACP, the Major City Chiefs' Association (MCCA), and the Major County Sheriffs' Association (MCSA).

The final note in this retrospective is arguably the most dramatic and important single step forward, one taken during this panel's period of work and around the time of the release of our Report 1. Acknowledging the inadequacy of UCR Summary-only count data to address important questions of interest, the FBI publicly announced its intent to shutter the UCR summary system by 2021 in favor of full NIBRS-format reporting. Formally, the mechanism for this "sunset" of the SRS was the approval by FBI Director James Comey on February 9, 2016, of a recommendation from the APB:

> The FBI UCR Program will transition to a NIBRS-only data collection by January 1, 2021, and will evaluate the probability of achieving that goal on an annual basis. Federal, state, local, and tribal agencies unable to meet the five year transition and who have committed to transitioning to NIBRS will collaborate with the FBI CJIS to develop a transition plan and timeline for conversion.

The decision was endorsed in a joint statement of the IACP, MCCA, National Sheriffs' Association, and MCSA. The imminent phaseout of the decades-old Summary Reporting System was announced in the most recent releases of UCR data, and the state UCR programs that coordinate inputs to the national level were formally apprised of the change in a June 10, 2016, letter from Director Comey.

The UCR Summary sunset is the most fundamental plank—but not the only one—in an ongoing UCR modernization program in process at the FBI. As our panel began its work, the UCR program was working to upgrade its computational infrastructure and break the constraints of having to work within the confines of processing and tabulation routines then still being performed on a 1960s-vintage computer mainframe housed in the FBI J. Edgar Hoover Building. (The UCR program only switched away from permitting paper returns in 2014.) In a foreword message to the 2014 release of *Crime in the United States*, then-Director Comey alluded to another major improvement being made in the program: 27 years after enactment of the Uniform Federal Crime Reporting Act mandated that federal law enforcement agencies report offenses to the UCR Program, there would finally be a concerted effort to get them—including the FBI itself and its field offices—to do so. As Director Comey closed his note:

> Most worthwhile changes involve growing pains. Even though we are taking steps to minimize the anticipated bumps and bruises, this undertaking will require tremendous effort and resources by our law enforcement and government partners, as well as the understanding of the media and the public. But to continue in our current system without comprehensive data only stalls meaningful conversation and fuels empty debates, both within law enforcement and in the communities we serve. Developing the UCR Program into a collaborative effort that gives all of us useful information and clarity can tell us more about where we may have problems and how we can improve.

– C –

Coverage of Recommended Crime Classification in Current Crime Statistics

In this appendix, we illustrate the degree to which offenses in the panel's recommended classification of crime (Section 5.2.1 and Appendix D in Report 1 for the table of codes, Section 5.2.2 in Report 1 for the corresponding attributes) is covered by the Uniform Crime Reporting (UCR) Program and the National Crime Victimization Survey. Section C.1 breaks the classification into small tabular pieces, each of which is accompanied by a short narrative—explaining the rationale for scoring the match (or non-match) to the current national crime statistics, clarifying the definitions or concepts used in the current systems, or describing the possible coverage of an offense by other data sources, as appropriate. As described in Section 2.1, the labels we apply in characterizing the match are rough assessments: a ✓ indicating that the definitions or concepts are in general agreement (and so whether the data source is capable of generating estimate counts or measures of the specific offense, a ✕ indicating a general mismatch or lack of coverage, and parenthetical comments (expended upon in the narrative) suggesting caveats and major differences in concepts.

Our proposed classification in Report 1 includes, as an essential complement to the listing of categories, a list of attributes or contextual variables that we suggest as a minimal set of data items that should be collected for applicable offenses. Accordingly, Section C.2 of this appendix is a narrative concerning data elements that are commonly collected in incident-based crime reporting systems—both NIBRS and selected state crime databases for which documentation is readily available.

In both sections, reference to coverage in the UCR emphasizes coverage in the National Incident-Based Reporting System (NIBRS) component of the UCR Program, listing the appropriate NIBRS offense code or data element. Relevant details of coverage of the offense type in the older, long-standing UCR Summary Reporting System (denoted SRS or UCR Summary) are described in the narratives when applicable. When not directly cited, the details on how an offense is covered in NIBRS or Summary is derived from the most recent NIBRS (Federal Bureau of Investigation, 2015) or Summary (Federal Bureau of Investigation, 2013) user manual.

C.1 CORRESPONDENCE OF OFFENSE CATEGORIES TO CURRENT CRIME STATISTICS

In the interest of brevity, the tables in this appendix do not drill down to the finest level of subcategorization in our classification (the *X.X.X.X* level). Likewise, we omit some categories and noncategories that are "other acts not already mentioned"/remainder categories that serve the purpose of making the classification exhaustive.

1—Acts Leading to Death or Intending to Cause Death

Offense Code	In UCR/NIBRS?	In NCVS?
1.1 Murder and intentional homicide	✓ (09A murder and nonnegligent manslaughter)	✗
1.2 Nonintentional homicide	✓ (as 09A)	✗
1.2.1 Nonnegligent manslaughter	✓ (as 09A, combined with murder)	✗
1.2.2 Negligent manslaughter	✓ (09B negligent manslaughter)	✗

In both the NIBRS and UCR Summary, murder and nonnegligent manslaughter are combined in a single, inseparable category. Negligent manslaughter is set apart as a separate category, mainly for purposes of applying the Hierarchy Rule in the Summary system.[1]

[1]It is worth noting that our proposed classification (focusing on behaviors) attempts to keep the difference between types of manslaughter relatively simple, defining "negligent manslaughter" as "unintended death as a result of a negligent, reckless, or involuntary act that is not directed against the victim." Gradations of manslaughter in penal/criminal code language have led to more evocative

Homicide is unique in the UCR framework in that it is the lone offense for which the UCR program requests detailed incident-based reporting even for jurisdictions that report only UCR Summary tallies. This is done through filing a Supplementary Homicide Report (SHR) for each homicide. We make reference to SHR codes for some of the contextual, incident-specific information requested in the SHR in Section C.2 of this appendix. However, we display the SHR codes for circumstance of homicide here, for reference, in Box C.1.

Being premised upon interviews with crime victims about their experiences, crimes involving death of the victim are necessarily out of scope for victim surveys such as the NCVS.

Offense Code	In UCR/NIBRS?	In NCVS?
1.3 Assisting or instigating suicide	✕ (unless counted as murder)	✕
1.3.1 Unlawful assisted suicide	✕ (unless counted as murder)	✕
1.4 Unlawful euthanasia	✕ (unless counted as murder)	✕
1.5 Unlawful feticide	✕ (unless counted as murder of victim less than 1 year old)	✕
1.6 Unlawful killing associated with armed conflict	✕	✕

The other types of acts causing death of the victim are not covered by the current national crime data systems—and, indeed, are ill-defined as federal crimes. Assisted suicide and unlawful euthanasia are not covered by federal criminal law, and their legalization varies by state statute. With respect to unlawful feticide, abortion is deemed legal in the United States pursuant to Supreme Court rulings and so not covered in federal criminal law. (Appropriations acts have routinely prohibited federal funding of acts such as assisted suicide or abortion.) That said, restrictions on abortion do exist in state statutes, and vary markedly across them—for instance, restriction by fetal age or rendering abortion unlawful for minors without parental notification.

but complex distinctions. For instance, the Uniform Code of Military Justice distinguishes between "voluntary manslaughter" (killing, with "intent to kill or inflict great bodily harm" "in the heat of sudden passion caused by adequate provocation") and "involuntary manslaughter," the latter including killing "by culpable negligence" (10 USC § 919).

Box C.1 Circumstances of Homicide Coded in Supplementary Homicide Reports (SHR)

2	Rape	44	Argument over money or property
3	Robbery	45	Other arguments
5	Burglary	46	Gangland killings
6	Larceny	47	Juvenile gang killings
7	Motor vehicle theft	48	Institutional killings
9	Arson	49	Sniper attack
10	Prostitution and commercialized vice	50	Victim shot in hunting accident
17	Other sex offense	51	Gun-cleaning death—other than self
18	Narcotic drug laws	52	Children playing with gun
19	Gambling	53	Other negligent handling of gun
26	Other—not specified	59	All other manslaughter by negligence
32	Abortion	60	Other
40	Lovers triangle	70	All suspected felony type
41	Child killed by babysitter	80	Felon killed by private citizen
42	Brawl due to influence of alcohol	81	Felon killed by police
43	Brawl due to influence of narcotics	99	Circumstances undetermined

NOTES: A further subcircumstance is coded for circumstances 80–81, referring to justifiable homicide. The subcircumstance is based on whether the felon attacked a police officer or a civilian, whether they attempted flight from a crime or resisted arrest, or whether they were killed in the commission of a crime. For all SHR entries, agencies directly code a "situation" based on whether single/multiple/unknown victims and single/multiple/unknown offenders are involved.

SOURCE: Federal Bureau of Investigation (2012), which is the codebook for the most recent Inter-University Consortium for Political and Social Research (ICPSR) archive of SHR data. The actual SHR instrument calls for free-text response in a table column bearing the heading "Circumstances (Victim shot by robber, robbery victim shot robber, killed by patron during barroom brawl, etc.)." These codes are eventually derived from those text responses.

Under the Uniform Code of Military Justice, unlawful killing associated with armed conflict (which would include unlawful killing "without justification or excuse") would be considered as "murder" (10 USC § 918)—hence, a separate "crosswalk" to the specific subcategory would be required when Defense Incident-Based Reporting System (DIBRS) data are worked into national crime statistics collection.

2—Acts Causing Harm or Intending to Cause Harm to the Person

Offense Code	In UCR/NIBRS?	In NCVS?
2.1 Assault	✓ (albeit conceptualized differently)	✓ (as in UCR/NIBRS)
2.1.1 Serious assault involving shooting or discharge of a firearm	✓ (part of 13A aggravated assault)	✓
2.1.2 Serious assault by means other than discharge of a firearm	✓ (part of 13A aggravated assault)	✓
2.1.3 Minor assault	✓ 13B simple assault	✓
2.2 Threat	✓ (in part)	✓ (in part)
2.2.1 Serious threat through shooting or discharge of a firearm	✓ (part of/inseparable from 13A aggravated assault)	✓ (though extent of threat/number of other victims unknown)
2.2.2 Serious threat through the display or pointing of a firearm	✓ (part of/inseparable from 13A aggravated assault)	✓ (though extent of threat/number of other victims unknown)
2.2.3 Serious threat by means other than firearm	✓ (13C intimidation)	✓ (though extent of threat/number of other victims unknown)
2.2.4 Minor threat	✗ (unless deemed intimidation)	✓

The 2.1–2.2 offenses in our recommended classification are the means by which we restructure the UCR Program's original concept of "aggravated assault" (as distinct from other or "simple assault"). Accordingly, the correspondence between the classification and existing data series is off, by design. That counts of UCR/NIBRS "aggravated assaults" are highly publicized (yet still fall short of an accurate, comparative measure of shooting incidents) while "simple assaults" are subject only to far-less-scrutinized, arrest-only collection is a longstanding flaw in national crime statistics—and one in particular need of remedy.

NIBRS defines intimidation (13C) as threatening words or behavior that "place another person in reasonable fear of bodily harm … but without displaying a weapon or subjecting the victim to actual physical attack." However, NIBRS renders intimidation as a subset of assault; our recommended definitions use the actual inflicting of harm as the distinction between assault and threat.

Offense Code		In UCR/NIBRS?	In NCVS?
2.3	Acts against liberty	To a degree	✗
2.3.1	Abduction of a minor[2]	✓ (in NIBRS, as 100 kidnapping/abduction)	✗[3]
2.3.2	Kidnapping for ransom	✓ (as kidnapping/abduction)	✗
2.3.3	Illegal restraint	✗ (unless coded as human trafficking or kidnapping/abduction)	✗
2.3.4	Hijacking	✗	✗
2.3.5	Illegal adoption	✗	✗
2.3.6	Forced marriage	✗	✗

Mentioned as a candidate for possible addition as a major crime category in both the 1929 Uniform Crime Reporting handbook that launched the program and in the 1958 expert-panel assessment, kidnapping/abduction never made it into the Summary inventory of crimes. Instead, it is listed as an example of "All Other Offenses," the Part II, arrests-only catch-all category. Kidnapping/abduction won inclusion in NIBRS as category 100. Our recommended behavioral definitions are such that acts of 2.3.3 illegal restraint might be labeled as either human trafficking or kidnapping/abduction in NIBRS.

2.3.4 Hijacking is handled unusually in UCR protocols. Under the Summary system's rules and constrained choices, the closest analogue (and how it may be counted) is robbery—a violent act against persons (vehicle operator and occupants) that involves the theft of property (the vehicle). NIBRS directs that "Hijacking-Air Piracy" be "classif[ied] as substantive offense, e.g., Kidnapping/Abduction or Robbery" (Federal Bureau of Investigation, 2015:54).

Offense Code		In UCR/NIBRS?	In NCVS?
2.4	Slavery and exploitation	To a degree	✗[4]

[2]For brevity, this table omits subcategories 2.3.1.1–2.3.1.4: parental abduction, abduction by a family member, abduction by a legal guardian, and abduction by another person

[3]Though the NCVS has not been used to estimate abduction of minors, the periodic National Incidence Studies of Missing, Abducted, Runaway, and Throwaway Children (NISMART) in 1988 and 1999 involved use of survey methods.

[4]It should also be noted that household survey methods are unlikely to reach the "target" group of victims of both 2.4 and 2.5 offenses.

Offense Code		In UCR/NIBRS?	In NCVS?
2.4.1	Slavery and involuntary servitude	✕ (not as such; inseparable from 64A human trafficking, involuntary servitude)	✕
2.4.2	Forced labor[5]	✕ (not as such)	✕

Our recommended classification attempts to decouple component behaviors/actions with the broader bundle of offenses added to the UCR program as "human trafficking" offenses. Specifically, we restrict the mechanics/logistics of trafficking to the trafficking offenses defined under category 2.5, and would treat the actual subjecting of another person to forced labor/servitude as either a 2.4.1 or 2.4.2 offense. However, the UCR categorization treats "human trafficking, involuntary servitude" as a unitary offense (Part I in Summary, 64A in NIBRS) that fuses both pieces of the underlying activity that we decompose separately.

Offense Code		In UCR/NIBRS?	In NCVS?
2.5	Trafficking in persons	To a degree	✕
2.5.1	Trafficking in persons for sexual exploitation	✓ (64B human trafficking, commercial sex acts)	✕
2.5.2	Trafficking in persons for forced labor or services	✓ (64A human trafficking, involuntary servitude)	✕
2.5.3	Trafficking in persons for organ removal	✕	✕
2.5.4	Trafficking in persons for other purposes	✕	✕

For UCR purposes (both Summary and NIBRS), human trafficking offenses are differentiated between those for "involuntary servitude" and those for "commercial sex acts"(64A and 64B). As described above, our classification addresses the mechanics/facilitation side of trafficking in these 2.5-group offenses and the direct subjection-to-act as either 2.4 slavery and exploitation or 3.4/3.5 sexual exploitation.

[5]For brevity, this table omits subcategories 2.4.2.1–2.4.2.3: forced labor for domestic services, forced labor for industry services, and other forced labor.

Offense Code	In UCR/NIBRS?	In NCVS?
2.6 Coercion	✓	✕
2.6.1 Extortion or blackmail	✓ (210 extortion/blackmail)	✕
2.7 Negligent acts	✕ (unless coded as assault, murder, or other offense)	✕
2.7.1 Negligence in situations of persons under care	✕	✕
2.7.2 Professional negligence	✕	✕
2.7.3 Negligence related to driving a vehicle	✕	✕
2.8 Dangerous acts		✕
2.8.1 Acts that endanger health of another person	✕ (not as such; some such actions could be coded as 90Z All Other Offenses)	✕
2.8.2 Operation of a vehicle under the influence of alcohol or other psychoactive substances	✓ (albeit Group B/arrests only)	✕

Two offenses in this range—extortion/blackmail and driving under the influence—were added to NIBRS when it began operation, albeit the latter only as a Group B offense for which arrests are tallied. General acts of criminal negligence (2.7) are not counted as such in the UCR save to the extent that they are part of other offenses, such as negligent manslaughter. Some of the examples given in Report 1 of 2.8.1 acts that endanger health of another person, such as violations of health and safety laws or adulterated foods/drugs/cosmetics, are listed in NIBRS documentation as part of Group B "all other offenses."

Offense Code	In UCR/NIBRS?	In NCVS?
2.9 Acts intended to induce fear or emotional distress	✕	✕
2.9.1 Harassment	✕ (only as part of 90Z all other offenses, arrests only)[6]	✕[7]
2.9.2 Stalking	✕ (not as such; defined in NIBRS as a example of 13C intimidation)	✓ (through supplemental survey)

[6]Harassment and 2.10 defamation are explicitly listed as sample offenses in the catch-all, Group B all-other-offenses category in the NIBRS manual.

[7]Individual instances of threatening behavior, perhaps not (yet) constituting a course of conduct, may be recorded in the NCVS.

Offense Code	In UCR/NIBRS?	In NCVS?
2.10 Defamation	✗ (not as such; only as part of 90Z all other offenses)	✗
2.11 Discrimination	✗	✗
2.12 Acts that trespass against the person	✗	✗
2.12.1 Invasion of privacy	✗	✗

The core offenses of category 2.9 are both referenced principally by example in the UCR Summary and NIBRS systems. In Summary, 2.9.1 harassment is not mentioned but 2.9.2 stalking is a defined example under other (simple) assault—and so subject to Part II, arrest-count-only coverage. In NIBRS, stalking is explicitly listed as an example of NIBRS's 13C intimidation offense, though it is not uncertain (and it seems unlikely) that NIBRS deals with either harassment or stalking as a "course of conduct" (pattern of two or more acts demonstrating a continuity of purpose) as our behavioral definition does. Instead, both 2.9.1 harassment and 2.10 defamation are listed as examples only in NIBRS's Group B "all other offenses" category.

3—Injurious Acts of a Sexual Nature

Offense Code	In UCR/NIBRS?	In NCVS?
3.1 Rape	✓ (statutory rape and sodomy separately defined)	✓ (but with less specificity than represented in these categories)
3.1.1 Rape with force	✓ (part of 11A rape [except statutory rape])	✓
3.1.2 Rape without force	✓ (part of 11A rape [except statutory rape])	✓ (except that reason/means not directly collected)
3.1.3 Rape involving inability to express consent or nonconsent	✓ (in part, as 36B statutory rape)	✓ (for age of nonconsent issues); — (for mental disability or other factors)
3.1.4 Threat of rape	✓ (part of 11A rape [except statutory rape])	✓

Offense Code	In UCR/NIBRS?	In NCVS?
3.2 Sexual assault	✗ (not as such; different concepts defined, as 11B sodomy, 11C sexual assault with an object, 11D fondling, 36A incest)	✓
3.2.1 Physical sexual assault	✗ (part of 11C sexual assault with an object, or other offenses above)	✓
3.2.2 Threat of a sexual nature	✗ (except as part of other offenses above)	✓

The original 1929 *Uniform Crime Reporting* manual defined rape as "the carnal knowledge[8] of a female, forcibly and against her will" (International Association of Chiefs of Police, 1929:195). This limited definition—excluding the rape of males and omitting a broad range of sexual assault offenses—remained completely unchanged for decades, with the 2004 UCR manual using exactly the same words and omitting only the comma after "female" (Federal Bureau of Investigation, 2004:19). In the UCR Summary System, "forcible rape" was partitioned into two categories, 2a rape by force and 2b attempts to commit forcible rape, with anything else being categorized as the Part II, arrests-only-reporting "other sex offenses." The *Blueprint for the Future of the Uniform Crime Reporting Program* argued for a broadening of the rape category to include "all forcible sexual offenses," including but not limited to "rape by instrumentation, rape of males, and other sexual assaults" (Poggio et al., 1985). When NIBRS was created out of the *Blueprint*, this definitional guidance was only partially adopted; the central "rape" definition was left unchanged, but a few related categories were created—the "rape by instrumentation" example was modified into 11C sexual assault with an object, and categories 11B sodomy and 11D fondling were added alongside "forcible rape" in NIBRS.

Even in 1929, UCR designers recognized the complexity of issues involving issues of consent, noting that their definition of rape "is subject to important extensions. [If] consent is given through fear of bodily harm, or the female

[8]Describing the continued use of the term in 2004, the *UCR Handbook* relied on the *Black's Law Dictionary*, 6th edition, definition of "carnal knowledge" as "the act of a man having sexual bodily connections with a woman; sexual intercourse" (Federal Bureau of Investigation, 2004:19), adding that "there is carnal knowledge if there is the slightest penetration of the sexual organ of the female (vagina) by the sexual organ of the male (penis)." By the 9th and 10th editions, *Black's* acknowledged that the term was "archaic" and redefined it as "sexual intercourse, esp. with an underage female."

is too young to be considered capable of consenting, or is of unsound mind, or is prevented from resisting by narcotics, etc., it is the policy of the law to punish for rape. The same is true of consent obtained by fraud or because of mistaken identity" (International Association of Chiefs of Police, 1929:195). However, the Summary system maintained its single definition, instructing participating agencies that they "must **not** classify statutory rape, incest, or other sex offenses" as forcible rape (unless the occurrence of statutory rape or incest involved "the female victim [being] forced against her will to engage in sexual intercourse" (Federal Bureau of Investigation, 2004:20). NIBRS made some concessions to these concerns, defining a separate category for 36B statutory rape to cover offenses by which consent could not be granted for sexual behavior due to age.

After years of continued pressure to expand the definition, the FBI finally modified the base definition of rape in 2011. The word "forcible" was stricken from category 2 rape, divided into 2a completed rape and 2b attempts to commit rape, and rape itself was now defined as "penetration, no matter how slight, of the vagina or anus with any body part or object, or oral penetration by a sex organ of another person, without the consent of the victim" (Federal Bureau of Investigation, 2013:13). For purposes of transition between the old and new definitions, a new subcategory 2c "historical rape" was established to permit agencies who could not immediately accommodate the new definition to continue to report under the old.

Offense Code	In UCR/NIBRS?	In NCVS?
3.3 Sexual violations of a nonphysical nature	✕	✕
3.4 Sexual exploitation of adults	✓ (to a degree, as 40A–C prostitution offenses)	✕
3.5 Sexual exploitation of children	✕ (not as separate category)	✕
3.5.1 Child pornography	✕	✕
3.5.2 Child prostitution, production and provision	✕	✕
3.5.3 Child prostitution, procurement	✕	✕

UCR Summary includes a Part II category for general "sex offenses (except rape and prostitution and commercialized vice)," which would be subject to arrests-only reporting and might be argued to include the offense of sexual exploitation of adults. In NIBRS, three separate categories (40A–C) are used for prostitution, assisting or promoting prostitution, and purchasing prostitution, respectively. Our suggested classification distributes various aspects of prostitution/sexual exploitation involving adults across several categories, somewhat differently. The central, exploitative, "pimping" acts are covered in category 3.4; the logistics of facilitating prostitution are considered trafficking in persons for sexual exploitation (2.5.1); and the procurement of prostitution is a prostitution offense under societal crimes (category 8.2.1).

4—Acts of Violence or Threatened Violence Against a Person That Involve Property

Offense Code	In UCR/NIBRS?	In NCVS?
4.1 Robbery	✓ (as general category, 120 robbery; all subtypes below fall under heading but are not differentiated)	✓ (in large part)
4.1.1 Robbery from the person	✓ (part of 120 robbery)	✓
4.1.2 Carjacking/robbery of a car or vehicle	✓ (part of 120 robbery)	✓
4.1.3 Robbery of valuables or goods in transit	✓ (part of 120 robbery)	✓ (but only in noncommercial settings, as a household survey)
4.1.4 Robbery of an establishment or institution	✓ (part of 120 robbery)	✕ (as a household survey)
4.1.5 Robbery of livestock	✓ (part of 120 robbery)	✕ (unless a household-based farm)
4.2 Terroristic or disruptive threats to buildings or critical infrastructure	✕	✕

Robbery is unique, conceptually and definitionally, in its fusion of elements of interpersonal violent crime with property crime. Accordingly, it makes sense that the offense would be addressed in a somewhat special manner in the UCR (and by the FBI, generally):

- Robbery was among the first offenses for which additional detail was sought even in UCR Summary counts, beginning in the early 1930s and continuing into the current Supplement to Return A, which UCR participants are requested to file each month alongside the usual counts. Box C.2 describes the content of the Supplement. Hence, in the UCR Summary, there is only one category for robbery—but the Supplement enables generation of a count

Box C.2 Counts and Summary Statistics Required by Monthly Supplement to Return A, Uniform Crime Reporting Program

Sums of estimated monetary value of property stolen and of property recovered by authorities, by type:

- Currency, notes, etc.
- Jewelry and precious metals
- Clothing and furs
- Locally stolen motor vehicles
- Office equipment
- Televisions, radios, stereos, etc.
- Firearms
- Household goods
- Consumable goods
- Livestock
- Miscellaneous

Number of "actual offenses" and sum of estimated monetary value of property stolen, in the course of all incidents of:

- Murder and nonnegligent manslaughter
- Rape
- Robbery, further subdivided by:
  - Highway (streets, alleys, etc.)
  - Commercial house
  - Gas or service station
  - Convenience store
  - Residence (anywhere on premises)
  - Bank
  - Miscellaneous
- Burglary/breaking or entering, further subdivided by:
  - Residence (dwelling), by night (6pm–6am), day (6am–6pm), or unknown
  - Non-residence (store, office, etc.), by night (6pm–6am), day (6am–6pm), or unknown
- Larceny-theft (except motor vehicle theft), further subdivided by:
  - $200 and over
  - $50 to $199
  - Under $50
- Motor vehicle theft (including "alleged joy ride")

Number of actual offenses and sum of estimated monetary value of property stolen, for all larceny-thefts (except motor vehicle thefts) further subdivided by:

- Pocket-picking
- Purse-snatching
- Shoplifting
- From motor vehicles (except parts and accessories)
- Motor vehicle parts and accessories
- Bicycles
- From building (except shoplifting and from coin-operated machines)
- From any coin-operated machines (parking meters, etc.)
- All other

Number of actual offenses involving motor vehicles recovered:

- Stolen locally and recovered locally
- Stolen locally and recovered by other jurisdictions
- Stolen in other jurisdictions and recovered locally

of "actual offenses" and a sum of estimated monetary value of stolen property for robbery perpetrated in seven general location types, including banks and residences. (The Supplement also requests estimates of the total monetary value of property stolen and recovered by law enforcement, by several general classes of property type, but this is not subdivided by the underlying offense, such as robbery or burglary.)

- Likewise, NIBRS defines only a single robbery category, but additional breakdowns by location type or property type may be derived from the other NIBRS variables. Our proposed classification differs by attempting to split out some variants of robbery by general mode, distinguishing (for example) carjacking—the stealing of a vehicle through direct harm or threatened harm of the driver—as materially different from other types of robbery.
- Bank robbery, as a specific variant of robbery, is especially curious from the standpoint of national crime statistics because it is subject to direct reporting—by the FBI—completely parallel to the UCR Program, and in that capacity has been the subject of more-than-just-total-count reporting for decades. Further, 18 USC § 2113 makes it a federal crime to rob any financial institution whose deposits are insured by the Federal Deposit Insurance Corporation (FDIC) or that is a member bank of the Federal Reserve—which is to say that the FBI has technical jurisdiction over most bank robberies, though investigations are typically done by local law enforcement (with the FBI taking a concurrent role). Annually, the FBI produces a Bank Crime Statistics (BCS) document (https://www.fbi.gov/investigate/violent-crime/bank-robbery/bank-crime-reports) independent of the UCR and *Crime in the United States*. The short report is a series of univariate analyses, presenting counts of bank robberies disaggregated by general type of institution (e.g., commercial banks or credit unions), age/sex/race of perpetrator, day of week, time of day, urbanization of the area, type of security devices at victimized institutions (and whether they were actually used/deployed), whether hostages were taken or not, and several other variables. The finest level of geographic detail in the report is a tabulation of total incidents by state. Like the UCR, the FBI's separate BCS tabulations leave considerable uncertainty about the quality and completeness of the information. To wit, in almost back-to-back statements at the end of each document, an implied claim of completeness:

  > The BCS provides a nationwide view of bank robbery crimes based on statistics contributed by FBI field offices responding to bank robberies or otherwise gathered when provided to the FBI from local and state law enforcement.

  is undercut by an added "Note:"

Not all bank robberies are reported to the FBI, and therefore BCS is not a complete statistical compilation of all bank robberies that occur in the United States.

As discussed in Report 1 (at p. 150), 4.2 terroristic or disruptive threats to buildings or critical infrastructure is an offense that we added because of its emergence in some states' criminal codes, and because it is similar to robbery as a hybrid violent/property crime (the threat of harm to individual persons through implied attack on "property").

5—Acts Against Property Only

Offense Code	In UCR/NIBRS?	In NCVS?
5.1 Burglary	✓ (220 burglary/breaking and entering)	✓ (but only household/person victims)
5.1.1 Burglary of business premises	✓	✗
5.1.2 Burglary of residential/private premises	✓	✓
5.1.3 Burglary of public premises	✓	✗

Burglary is handled much the same way in the UCR Summary and NIBRS as is robbery. Through the Supplement to Return A (Box C.2), the Summary system requests slightly more than an overall count; it requests an "actual offense" count and estimated monetary value in six categories—residence/non-residence crossed by time of day (night, day, or unknown). And NIBRS generates the same information through a single 220 burglary/breaking and entering code and the values of other data elements.

Burglary warrants special mention in terms of measurement issues because it occasioned one of relatively few formal counting rules in the UCR data systems. Known as the "Hotel Rule," the rule applies to burglaries of "hotels, motels, lodging houses, or other places where lodging of transients is the main purpose," as well as rental units in self-storage buildings (Federal Bureau of Investigation, 2015:25)—in short, how to count burglary offenses in situations in which multiple rooms or units within a structure were likely violated in a single instance. Under the "Hotel Rule," only a single incident would be logged in NIBRS in places like hotels where the individual units are under the supervision of a single manager and "[that] manager, rather than the individual tenants/renters, will most likely report the offenses to the police." Within that single incident, the "number of premises entered" is still entered as a NIBRS data element. But, if both those conditions fail—"if multiple occupants rent or lease individual living or working areas in a building for a period of time, which would preclude the tenancy from being classified as transient, and the occupants would most likely report the individual burglaries separately"—then the individual burglaries are to be reported in NIBRS as separate incidents.

Offense Code		In UCR/NIBRS?	In NCVS?
5.2	Theft	✓	✓ (but only household/person victims)
5.2.1	Theft of a motorized vehicle or parts thereof	✓ (albeit as separate categories, 240 motor vehicle theft and 23G theft of motor vehicle parts or accessories)	✓
5.2.2	Theft of personal property	✓ (though subdivided differently in 23A–H, e.g., pocket-picking, theft from motor vehicles, theft from building)	✓
5.2.3	Theft from business or other nonpublic organization	✓ (if reported to law enforcement)	✕
5.2.4	Theft of public property	✓	✕
5.2.5	Theft of livestock	✓	✕ (unless household-based farm)
5.2.6	Theft of services	✓ (in principle, part of 23H all other larceny)	✕

Via the Supplement to Return A (Box C.2), the UCR Summary system divides the top-level larceny-theft category (except motor vehicle theft, which is itself deemed a separate Part I offense) into 9 subcategories, for which the number of "actual offenses" and estimated monetary value of stolen property are meant to be summed. In order, the subcategories are: pocket-picking, purse-snatching, shoplifting, theft from building, theft from coin-operated device/machine, theft from motor vehicles, theft of motor vehicle parts/accessories, theft of bicycles, and all other larceny. Larceny-theft is the exception to a general rule in that it is one offense for which the built-in categories in the Summary system are more detailed than those in NIBRS: NIBRS combines the "theft of bicycles" subcategory with all other larceny. (Of course, the detail on theft of bicycles relative to other types of property can be recovered from the NIBRS data element on type of property involved in the offense.)

Motor vehicle theft is a separate Part I offense, and the Supplement to Return A asks for counts (but not estimated monetary values) to be tallied separately for three different classes—for motor vehicle thefts in which the vehicle is recovered after the fact. The three-way split is based on whether the vehicle was stolen in the reporting department's jurisdiction (or not) or recovered in the reporting department's jurisdiction (or not).

Offense Code		In UCR/NIBRS?	In NCVS?
5.3	Acts against computer systems	✗	✗
5.3.1	Unlawful access to a computer system	✗/✓ (hacking and unauthorized access added to NIBRS in 2015; limited data as yet)	✗
5.3.2	Unlawful interference with a computer system or computer data	✗	✗
5.3.3	Unlawful interception or access of computer data	✗	✗
5.4	Intellectual property offenses	✗	✗

Computer-based crimes are not included in the UCR Summary save to the degree to which some might be coded in Part II's catch-all, arrest-only all other offense group. As with animal cruelty later in the classification, we both credit NIBRS with coverage and note 5.3.1 unlawful access to a computer system as a gap, as the closest match to the hacking/computer invasion offense that was formally added to NIBRS in 2015 but for which data has yet to be tabulated. Our classification differs with the NIBRS package by considering acts against computer systems a larger category in its own right rather than a subset of fraud offenses; it also adds the subcategory offenses of unlawful interference and unlawful interception or access of data (separate from breaches of computer systems).

Our classification system relies heavily on the cybercrime-involved flag included in the attributes section to indicate offenses where computers or networks are critical to the modus operandi of the criminal offense. It is in this way that we avoid carving out categories for specific cybercrime offenses for which keeping up with terminology could be a losing battle. To this end, we rely on the cybercrime flag to facilitate construction of derived categories of interest, such as ransomware (in terms of behavior, the combination of extortion and cybercrime-involvement) or cyberbullying (harassment in combination with cybercrime).

Category 5.4 intellectual property offenses are new to U.S. crime data collection.

Offense Code	In UCR/NIBRS?	In NCVS?
5.5 Property damage	✓ (290 destruction/damage/vandalism of property)	✗
5.5.1 Arson	✓ (200 arson)	✗
5.5.2 Reckless burning	✗	✗

Arson is an inherently difficult offense to measure or count, immediately, because investigation is typically required to determine whether damaging fire was "willfully" set (or, in NIBRS parlance, "unlawfully and intentionally set;" Federal Bureau of Investigation, 2015:21). Given the measurement difficulty, it is perhaps appropriate that arson may be the most fragmented of major offense types in terms of authority for data collection, with at least three nationally compiled databases all bearing a legitimate claim to be a measure of record:

- As noted in our Report 1, the UCR Program is not itself explicitly defined under federal law—yet the collection of data on arson by the UCR *is* so authorized. The UCR Program was directed by law in 1978 (and again in 1982) to consider arson a Part I offense (92 Stat. 3465; 96 Stat. 1319).
- Shortly before directing the addition of arson as a Part I UCR offense, Congress renamed and reauthorized the U.S. Fire Administration (USFA) to "formulate methods for collection of arson data" and to compile "nationwide arson statistics within the National Fire Incident Reporting System [NFIRS]" (92 Stat. 933). Now housed within the Federal Emergency Management Administration (FEMA), the USFA's NFIRS database is intended to be a repository of voluntary-report information on all fire department incidents, paralleling the role of the UCR Program for law enforcement agencies. NFIRS continues to include a dedicated Arson Module for completion in arson incidents, in addition to the report required for any incident of burning.
- In 1996, a provision in appropriations legislation authorized the Secretary of the Treasury to "establish a national repository of information on incidents involving arson and the suspected criminal misuse of explosives." Submitting records on such incidents was made mandatory for federal agencies (pursuant to subsequent regulations), but it was also noted that the repository would "contain information on incidents voluntarily reported to the Secretary by State and local authorities" (110 Stat. 3009–369; provision now codified at 18 USC § 846(b)). The Homeland Security Act of 2002 that organized the new Department of Homeland Security also transferred authority for this arson and explosives repository to the Attorney General (as part of transferring the Bureau of Alcohol, Tobacco, Firearms, and Explosives [ATF] to the Justice Department; P.L. 107-296). The repository created pursuant to this law is the Bomb Arson Tracking System (BATS) of the U.S. Bomb Data Center, which

itself is expected to be permanently co-located with ATF's National Center for Explosives Training and Research at Redstone Arsenal, Alabama.[9] A "frequently asked question" on the ATF's website describes the distinction between NFIRS and BATS as follows: "The primary mission of NFIRS is to collect fire incident information" while "BATS is dedicated to documenting the 'follow-up' investigation (i.e., case management)."[10] The distinction is not as clear as might be hoped, of course, because (as noted at the outset) investigation is typically needed to document whether an incident of arson has actually taken place.

For purposes of crime statistics and analysis, the offense of arson has traditionally been conceptualized as a pure property crime. In our classification as in the UCR, harm done to persons inside of a structure would be treated separately in offense counts (arson being an exception to the UCR's Hierarchy Rule for collapsing incidents down to a single offense). Compounding the difficulty of national/federal collection of data on arson is that federal law alone tends to blur the distinction between property and violent crime in the two major instance in which arson is defined. As a federal crime "within the special maritime and territorial jurisdiction of the United States," the offense of arson can be prosecuted differently (with different penalties applying) "if the life of any person be placed in jeopardy" (or even if the structure burned is simply a dwelling/residential location) (18 USC § 81).[11] Meanwhile, article 126 of the Uniform Code of Military Justice (UCMJ; 10 USC § 926) that would apply to arson committed by military personnel formally distinguishes "simple arson" from "aggravated arson" akin to the UCR concept of assault—with "aggravated arson" applying if "to the knowledge of the offender there is at the time [of the burning] a human being" inside the dwelling or structure.

The only other UCR/NIBRS offense that covers instances of property damage other than arson is the somewhat catch-all category 290 destruction/damage/vandalism of property. In our classification, we opted to downplay vandalism, which is typically viewed as a relatively petty crime, and fold it into category 5.5.3's "other property damage." That assessment

[9]This administrative transfer was referenced, and commended, by Senate appropriators in their explanatory statement accompanying the Commerce, Justice, and Science appropriations bill for fiscal year 2017 (S.Rept. 114-239). In particular, the appropriators noted that the Department of Justice "has designated the [BATS] repository maintained at USBDC as the sole repository for explosives incidents, and has directed all DOJ components to use this system to document explosives-related incidents. The Committee believes that the USBDC will be more productive with regular input and analysis from multiple agencies."

[10]See https://www.atf.gov/explosives/qa/isn't-bats-same-national-fire-incident-reporting-system-nfirs.

[11]The federal definition covers the willful and malicious burning of "any machinery or building materials or supplies, military or naval stores, munitions of war, or any structural aids or appliances for navigation or shipping," and attempt or conspiracy to commit arson is treated equivalently with the actual act of arson (18 USC § 81).

is certainly ripe for review, following the collection of more data and settling upon appropriate indicators of arson and burning.

6—Acts Involving Controlled Substances

Offense Code	In UCR/NIBRS?	In NCVS?
6.1 Unlawful possession or use of controlled drugs for personal consumption	✕ (conceptualized differently; 6.1–6.3 all may be part of generic 35A drug/narcotic violations)	✕
6.2 Unlawful cultivation or production of controlled drugs	✕	✕
6.3 Unlawful trafficking or distribution of controlled drugs	✕	✕
6.3.1 Street-level selling of quantities of controlled drugs suitable for personal consumption	✕	✕
6.3.2 Wholesale distribution/trading/possession of controlled drugs	✕	✕
6.4 Unlawful acts involving drug equipment or paraphernalia	✓ (35B drug equipment violations)	✕

As we describe in the text, category 6 on controlled substances is at the nexus of what could be considered "traditional" crime for purposes of crime statistics and what is "new" crime for data collection. As a criminogenic factor, drug involvement in crime, drug market-related crime, and illicit drug use are all commonly thought to be well known or well documented but on which national crime statistics are surprisingly silent. Drug involvement in other violent crimes has long been impossible to recover from UCR Summary data; meanwhile, a catch-all "drug abuse violations" is included in UCR Summary only as a Part II, arrests-only (and so relatively uninformative) category. NIBRS represented a slight improvement, crafting separate offense categories for general violations of drug/narcotic laws and violations involving possession or use of drug equipment (e.g., for production and resale of drugs). But the mismatch between the NIBRS "drug/narcotic violations" and the more market-based, possession-distinct-from-production-distinct-from-trafficking categories we suggest in our classification is sufficiently great that we think it fairest to consider the current data as a nonmatch. Substantively, our 6.4 unlawful acts involving drug equipment or paraphernalia is a different case in that it does generally match the NIBRS drug equipment offense—but, again, NIBRS's current limitation to arrests only potentially misses a large share of the drug-related offending that is of great public safety (and public health) importance.

7—Acts Involving Fraud, Deception, or Corruption

Offense Code	In UCR/NIBRS?	In NCVS?
7.1 Fraud	✓ (in parts; conceptualized very differently)	✗
7.1.1 Consumer financial and products/services fraud	✓ (most consistent with 26A false pretenses/swindle/confidence game, but also some of other NIBRS categories like 26B credit card/automated teller machine fraud or 26E wire fraud)	✗
7.1.2 Identity theft	✗/✓ (26F identity theft added in 2015, but no data available yet)	✓ (through topic supplement)
7.1.3 Fraud against businesses or establishments, including nonprofit organizations	✓ (could be part of 26B false pretenses/swindle/confidence game)	✗
7.1.4 Fraud against government agencies	✗ (not as such, though 26D welfare fraud would be special case)	✗

The UCR Summary system includes a single "fraud" line item as a Part II, arrests-only reporting offense—at best, a weak indicator for a class of offenses as harmful and as complex as fraud. NIBRS represented an improvement by adding numerous fraud-related categories. Arguably, the central one in the set is 26A false pretenses/swindle/confidence game, covering a wide range of criminally deceptive actions and culturally well-known concepts such as confidence games/scams. NIBRs also included several categories tailored more to the mechanics of fraud (e.g., credit card/automated teller machine fraud and wire fraud) than to distinct differences in the underlying behavior. Inasmuch as the UCR systems are generally tailored more toward individuals as the actors (victims and offenders), coverage of fraud against businesses or establishments in NIBRS is mainly tangential, and the specific NIBRS category of 26D welfare fraud—if detected and reported to local law enforcement—would stand as a special case of the broader class of fraud against government agencies.

Identity theft, along with animal cruelty and hacking/computer invasion, are the newest additions to the NIBRS repertoire, and so are credited here as both a match and a nonmatch on coverage relative to our classification. Identity theft is now a named NIBRS offense, but it is not one for which data have yet been widely promulgated. It should be noted that identity theft is an instance in which the NCVS shines in terms of its coverage, having been the subject of topic supplements to the main survey and so generated some of the most thorough (and usefully provocative) national statistics on an increasingly important offense type.

Offense Code		In UCR/NIBRS?	In NCVS?
7.2	Forgery/counterfeiting	✓ (250 forgery/counterfeiting, without subcategory detail below)	✗
7.2.1	Counterfeiting means of payment	✓	✗
7.2.2	Counterfeit product offenses	✓	✗
7.2.3	Acts of forgery/counterfeiting documents	✓	✗

UCR Summary defines forgery/counterfeiting as a single Part II offense, subject to arrests-only reporting. NIBRS likewise defines a single category for 250 counterfeiting/forgery. Our recommended classification defines a few subcategories, in that counterfeiting currency/means of payment seems sufficiently, substantively different from document forging as to warrant separate breaks, though it might be possible to reconstruct some of this information from the NIBRS code in conjunction with other NIBRS data elements.

Forgery/counterfeiting is interesting in the NIBRS context because a formal counting rule has developed around it—invoked if an underlying rationale for the counterfeiting is actually realized. Until a forged/counterfeit financial instrument (like a check) is actually passed and items are received in return, then the forgery/counterfeiting is the only reportable/countable offense in NIBRS. But, at the point that goods or something of value is obtained in return for the passed counterfeit instrument, then that instrument becomes "negotiable" in the financial sense—and NIBRS guidance is that both fraud and counterfeiting occurred and both should be recorded (Federal Bureau of Investigation, 2015:26).

Offense Code		In UCR/NIBRS?	In NCVS?
7.3	Corruption	✓ (in part)	✕
	7.3.1 Bribery	✓ (510 bribery)	✕
	7.3.2 Embezzlement	✓ (270 embezzlement)	✕
	7.3.3 Abuse of functions	✕	✕
	7.3.4 Trading in influence	✕	✕
7.4	Acts involving proceeds of crime	✓ (generally consistent with 280 stolen property offenses)	✕
	7.4.1 Money laundering	✓ (as part of 280)	✕
	7.4.2 Illicit trafficking in cultural property	✓ (as part of 280)	✕
	7.4.3 Fencing stolen goods	✓ (as part of 280)	✕

Several of the white-collar offenses in this block are defined within NIBRS, though the quality of the resulting data is largely unknown or unexplored. Of these, only embezzlement and "stolen property: buying, receiving, possessing" factor into the UCR Summary system—and then only as Part II, arrests-only offenses. NIBRS added a category for bribery and retained a general "stolen property offenses" category, into which general offenses related to proceeds of crime might be said to fall.

The offenses in this particular block are of additional interest because they are common examples of what are legally described as "racketeering activity"—individual occurrences of which would properly be counted in these and other substantive categories, but "patterns" of occurrence (course of conduct) would also properly trigger being counted as a racketeering offense/violation of the federal Racketeer Influenced and Corrupt Organizations (RICO) Act. See the discussion of organized criminal group offenses at 9.3 in the classification, below.

8—Acts Against Public Order and Authority

Offense Code		In UCR/NIBRS?	In NCVS?
8.1	Acts against public order behavioral standards	✕	✕
	8.1.1 Violent public disorder offenses	✕ (disorderly conduct is Group B, arrests-only reporting offense)	✕
	8.1.2 Acts related to social public order norms and standards	✕	✕

Offense Code	In UCR/NIBRS?	In NCVS?
8.2 Acts against public order sexual standards	✓ (in large part)[12]	✗
8.2.1 Prostitution offenses	✓ (40A–C, separate categories for prostitution and commercialized vice, assisting/promoting prostitution, and purchasing prostitution)	✗
8.2.2 Pornography offenses	✓ (370 pornography/obscene material)	✗
8.3 Acts related to freedom of expression or control of expression	✗	✗
8.3.1 Acts against freedom of expression	✗	✗
8.3.2 Acts related to expressions of controlled social beliefs and norms	✗	✗

This grouping of "public order" offenses is a mixed bag in terms of correspondence with existing UCR standards. The closest concordance, historically, is with 8.2.1 prostitution offenses; as discussed above regarding human trafficking for sexual exploitation, we intend 8.2.1 prostitution offenses to cover the purchase/client side of prostitution, which is consistent with UCR Summary's Part II concept of purchasing prostitution (arrests-only reporting) and NIBRS's 40C purchasing prostitution. NIBRS added a category for offenses related to pornography/obscene material, which was not otherwise addressed in UCR Summary. On public order behavior offenses, NIBRS's Group B, arrests-only disorderly conduct code loosely corresponds with part of 8.1.1 violent public disorder offenses.

Offense Code	In UCR/NIBRS?	In NCVS?
8.4 Acts contrary to public revenue or regulatory provisions	✗	✗
8.4.1 Tax evasion, and other acts against taxation provisions	✗	✗
8.4.2 Market manipulation, insider trading, and other acts against market or financial regulations	✗	✗

[12]NIBRS defines 36A incest as a subcategory of "sex offenses, nonforcible"; the panel's classification places incest under 8.2.3 other acts against public-order sexual standards.

Offense Code		In UCR/NIBRS?	In NCVS?
8.4.3	Acts against regulations on alcohol, tobacco, or gambling	✓ on gambling, — on alcohol/tobacco[13]	✗
8.4.4	Customs violations	✗	✗

The white-collar offenses in our categories 8.4.1 and 8.4.2 are new terrain for U.S. national crime statistics while—aptly, for a system born in the 1920s—the "sin tax"offenses bundled into category 8.4.3 are somewhat better covered. Generic categories for "gambling" and "liquor laws" are included as Part II, arrests-only categories in the UCR Summary. NIBRS retained the same treatment of alcohol, retaining "liquor law violations" in its Group B, arrests-only class. Gambling violations, by comparison, were elevated somewhat in NIBRS, with four separate subcategories delineated.

As discussed further under category 9.3 below, the "sin tax" provisions and NIBRS's subcategories for gambling offenses are among the types of offenses that may trigger the separate offense of racketeering violations when conducted more than once.

Offense Code		In UCR/NIBRS?	In NCVS?
8.5	Acts related to migration	✗	✗
8.5.1	Offenses related to smuggling of migrants	✗	✗
8.5.2	Unlawful entry/border crossing	✗	✗
8.5.3	Unlawful employment or housing of an undocumented migrant	✗	✗

Migration offenses are new terrain for U.S. national crime statistics (but are certainly increasingly important to try to understand). There are, inherently, concerns about possible overlap with human trafficking offenses (described earlier)—and these and other migration-related offenses are notable parts of the federal definition of "racketeering activity" (as discussed relative to category 9.3, below), but the definitions outlined in Report 1 are intended to demarcate the varying behaviors specific to each category.

[13]For gambling, the NIBRS defines separate subcategories for betting/wagering, operating/promoting/assisting gambling, gambling equipment violations, and sports tampering (39A–D). On alcohol, "liquor law violations" are denoted only as a Group B, arrests-only offense.

Offense Code		In UCR/NIBRS?	In NCVS?
8.6	Acts against the justice system	X	X
	8.6.1 Obstruction of justice	X	X
	8.6.2 Breach of justice system authority	X	X
	8.6.3 Preparatory or enabling crimes	X (not as such)	X

There is certainly a difference between "crime statistics" and the broad class of "justice statistics" that speak to the functioning of the entire criminal justice system; still, however paradoxically, offenses against the justice system itself are essentially uncovered in current crime statistics. Inchoate offenses—preparatory or enabling crimes—are only addressed in small part. To wit, several such inchoate offenses listed in NIBRS documentation are explicitly named examples of "offenses of general applicability" in NIBRS's Group B, arrests-only, 90Z All Other Offenses category. Specifically, Federal Bureau of Investigation (2015) holds that "Offenses of General Applicability are those offenses prefixed by 'Accessory Before/After the Fact,' 'Aiding and Abetting,' ['Conspiracy] to Commit,' 'Enticement,' 'Facilitation of,' 'Solicitation to Commit,' [or] any other [similar] prefix"—to be included in 90Z (if at all) "unless it is an integral component of [a] Group A offense such as human trafficking."

Offense Code		In UCR/NIBRS?	In NCVS?
8.7	Acts related to democratic elections	X (election law violations cited as example of 90Z all other offenses, Group B arrests-only reporting)	X
	8.7.1 Acts intended to unduly influence voters at elections	X	X
8.8	Acts contrary to labor law	X	X
	8.8.1 Collective labor law violations	X	X
	8.8.2 Individual labor law violations	X	X

On 8.7, "election law violations" is listed as an example of things that might be included in "all other offenses" (90Z), for Group B arrests-only reporting. Misuse of labor organization/union funds is part of a long list of possible "racketeering activities" (see 9.3 below), but actual infractions of labor law are new to crime statistics.

Offense Code	In UCR/NIBRS?	In NCVS?
8.9 Acts contrary to juvenile justice regulations or involving juveniles/minors	X (not as such)	X
8.9.1 Status offenses	X (not as such)	X

In the UCR Summary, violations by juveniles (persons under age 18) of local curfew or loitering laws is labeled a Part II, arrests-only offense. Historically, "runaways"—also applying to persons under age 18—was also a Part II offense, but "arrest data on Runaways is no longer required by the FBI UCR Program due to the differences between jurisdictional statutes" (Federal Bureau of Investigation, 2013:169). NIBRS is likewise fairly silent on juvenile justice statistics—deferring, perhaps to the parallel juvenile justice system. NIBRS does include a Group B, arrests-only category for curfew/loitering/vagrancy violations that is applicable to both juveniles and adults—youth curfew violations being a common example of a status offense.

9—Acts Against Public Safety and National Security

Offense Code	In UCR/NIBRS?	In NCVS?
9.1 Acts involving weapons, explosives, and other destructive materials	X (general "weapon law violations" is considered a Group B, arrests-only offense)	X
9.1.1 Unlawful possession or use of weapons and explosives	X	X
9.1.2 Trafficking of weapons and explosives	X	X
9.2 Acts against national security	X	X

These general public safety/national security violations are generally new to crime data collection. The exception, loosely, being violations of weapons laws—the unusually crafted "Weapons: Carrying, Possessing, Etc." category is a UCR Part II offense and a generic "Weapon Law Violations" is a NIBRS Group B offense, both subject to arrests-only reporting. Some of the named examples of acts against national security, including espionage, are among those listed in the NIBRS handbook as Group B, arrests-only 90Z all other offenses.

Offense Code	In UCR/NIBRS?	In NCVS?
9.3 Acts related to organized criminal groups	X (not as such)	X
9.3.1 Racketeering, and violations of the Racketeer Influenced and Corrupt Organizations (RICO) Act	X	X

NIBRS defines no separate category for racketeering—very loosely, the "business" side of organized crime, meaning travel or the use of mail or modes of transportation to further the ends of a criminal enterprise. Instead, NIBRS guidance on the treatment of RICO violations (Federal Bureau of Investigation, 2015:60) is to "report predicate offenses, e.g., Arson, Aggravated Assault, Extortion/Blackmail, or Human Trafficking"—the crimes enabled or facilitated by racketeering activity. In NIBRS, organized crime involvement can be coded as contextual information for selected offense types:

- As "Other Gang" (as opposed to juvenile gangs), in Gang Involvement variable applicable to NIBRS-defined counterfeiting/forgery, stolen property offenses, drug/narcotic violations, drug equipment violations, pornography/obscene material, gambling equipment violations, weapon law violations, and animal cruelty; or
- As "Gangland (Organized Crime Involvement)," in Aggravated Assault/Homicide Circumstances variable for those two offense types.

Our proposed classification includes a group support/motivation attribute to be flagged for individual offenses, consistent with the Gang Involvement variable in NIBRS, to make it possible to assess the extent of gang or other organized-enterprise involvement in particular types of offenses. But it also includes category 9.3 to properly consider racketeering activity as part of the overall picture of crime. Indeed, by the letter of the act, RICO violations are (like stalking and harassment) technically course-of-conduct offenses that require demonstration of multiple individual offenses (a pattern) to be triggered as offenses in their own right; see Box C.3 for additional detail. Such course-of-conduct offenses are challenging to handle in the usual data collections, but are nonetheless important and damaging offenses to understand.

Box C.3 Central Definitions in the Racketeer Influenced and Corrupt Organizations (RICO) Statute

As federal statute—codified as Title 18, Chapter 96 of the U.S. Code—the Racketeer Influenced and Corrupt Organizations (RICO) Act defines offenses relative to "patterns of racketeering activity" and "enterprises engaged in, or the activities of which affect, interstate and foreign commerce." Generally speaking, it is unlawful under RICO for the proceeds of such a pattern of racketeering activity to be applied to the benefit of such an enterprise; or for such a pattern to be used to acquire interest/control in such an enterprise; or for any person associated with such an enterprise to conduct the enterprise's work through such a pattern; or for any person to conspire along any of these lines (18 USC § 1962).

Accordingly, the definitions of "racketeering activity" and "pattern of racketeering activity" in 18 USC § 1961 are crucial. On the latter, the act holds that demonstrating a pattern "requires at least two acts of racketeering activity, [the] last of which occurred within ten years (excluding any period of imprisonment) after the commission of a prior act of racketeering activity." The definition of "racketeering activity" is extremely broad, and includes (excerpting, paraphrasing, and rearranging the capsule references to code sections included in the definition itself):

- **State-defined offenses:** "Any act or threat involving murder, kidnapping, gambling, arson, robbery, bribery, extortion, dealing in obscene matter, or dealing in a controlled substance" that is chargeable under relevant state laws "and punishable by imprisonment for more than one year" by state criminal code;
- **Offenses defined in federal criminal code (Title 18, U.S. Code):** "Any act which is indictable" under numerous sections of federal criminal code, including:
  - racketeering itself (travel or use of mail/transportation to commit violent crime in furtherance of other unlawful activity, distributing proceeds of unlawful activity, or generally promote or manage unlawful activity)
  - interference with commerce, as through robbery or extortion
  - money laundering, including transactions in property derived from unlawful activity
  - bribery, including sports bribery
  - counterfeiting, including trafficking in goods bearing counterfeit marks
  - felonious theft from interstate shipment
  - embezzlement from pension and welfare funds, including receipt of unlawful welfare fund payments
  - extortionate credit transactions
  - fraud related to identification documents and access devices (e.g., credit cards, bank account numbers)
  - mail fraud and wire fraud, including operation of unlawful money transmitter businesses
  - financial institution fraud in foreign labor contracting
  - economic espionage and theft of trade secrets
  - murder-for-hire, and use of interstate commerce in its commission
  - offenses against the justice system, including obstruction of justice (such as of criminal investigations or of state/local law enforcement) and tampering with or retaliating against a witness, victim, or informant
  - offenses involving naturalization or citizenship documents, including unlawful procurement or reproduction of the same, as well as forgery/false use/misuse of passports, visas, permits, or other such documents
  - peonage, slavery, and trafficking in persons, or the sexual exploitation of children

Box C.3 (continued)

- **Offenses defined in federal criminal code** (continued)
  - gambling offenses, including the unlawful transmission of gambling information, interstate transportation of wagering paraphernalia, or operation of illegal gambling businesses
  - transmission or distribution of obscene matter
  - interstate transportation of stolen property or stolen motor vehicles
  - criminal copyright infringement, including trafficking in counterfeit audiovisual works or computer software or trafficking in unauthorized recordings or videos of live musical performances
  - trafficking in motor vehicles or parts, for which the identification number has been removed, damaged, or altered
  - trafficking in contraband cigarettes
  - development, acquisition, possession, or trafficking in biological weapons, chemical weapons, or nuclear materials;
- **Embezzlement from or unlawful transactions involving union/labor organization funds**, as defined elsewhere federal tax code;
- **Additional types of fraud** defined in federal bankruptcy code (Title 11 of U.S. Code) or the fraudulent sale of securities, as well as **trafficking in controlled substances** ("the felonious manufacture, importation, receiving, concealment, buying, selling, or otherwise dealing in a controlled substance or listed chemical" as "punishable under any law of the United States");
- **Certain suspicious financial transactions**: Indictable offenses under the federal Currency and Foreign Transactions Reporting Act;
- **Additional indictable offenses related to immigration** (e.g., harboring of aliens or "importation of alien for immoral purpose") "if the [indictable act] was committed for the purpose of financial gain; and
- **Any act generally indictable as a federal terrorism offense**: A category that, itself, references numerous other offenses in federal criminal code, including such provisions as acts of violence against transportation or communication infrastructure, threats against federal government officials, and financing or providing material support of terrorism.

Offense Code	In UCR/NIBRS?	In NCVS?
9.4 Terrorism	X (not as such)	X
9.4.1 Participation in a terrorist group	X	X
9.4.2 Financing of terrorism	X	X

The handling of terrorism-related offenses in NIBRS is similar to racketeering/organized crime offenses (directly above), in that the NIBRS manual's guidance is to "classify as substantive offense," in order to count (as listed examples) "homicide or property destruction" involved with a terrorist incident. But, unlike the organized crime example, there is no provision for flagging terrorism involvement as an attribute of the substantive NIBRS offense, so that information is conceptually lost in aggregation—as are the participation/enabling offenses we define as subcategories in our classification. (We also include "terrorism-related" as a desired value in the group support/motivation attribute that is part of the classification.)

10—Acts Against the Natural Environment or Against Animals

Offense Code	In UCR/NIBRS?	In NCVS?
10.1 Acts that cause environmental pollution or degradation	X	X
10.2 Acts involving the movement or dumping of waste	X	X
10.3 Trade or possession of protected or prohibited species of fauna and flora	X	X
10.4 Acts that result in the depletion or degradation of natural resources	X	X
10.4.1 Illegal logging or mining	X	X
10.4.2 Illegal hunting, fishing, or gathering of wild fauna and flora	X	X

Environmental violations are major, new terrain for U.S. crime statistics—the only reference to offenses in this category in NIBRS materials being that "environmental law violations" is listed as an example of the NIBRS Group B 90Z "all other offenses" category.

Offense Code	In UCR/NIBRS?	In NCVS?
10.5 Acts against animals	✗/✓ (720 animal cruelty, recently added)	✗

Acts against animals (10.5) is listed as both a gap and a covered offense due to the newness of its addition to NIBRS. The UCR Program announced the addition of animal cruelty to NIBRS "as a Group A offense and as a Crime Against Society" in January 2015, with data collection to begin on January 1, 2016. In addition to counting the offense under code 720 (animal cruelty), four abuse types were made available for coding in NIBRS in the "Type [of] Criminal Activity/Gang Information" variable (Criminal Justice Information Services Division, 2015:9): simple/gross neglect, intentional abuse and torture, organized abuse (e.g., dogfighting, cockfighting), and animal sexual abuse (bestiality). Absent knowledge of the quality and completeness of data collected under the new code, we opted to keep the top-line "acts against animals" as a crime category without specific subcategory breaks, but the examples/inclusions we list in the detailed classification (Appendix D of Report 1) are consistent with the NIBRS-defined abuse types.

11—Other Criminal Acts Not Elsewhere Classified

Offense Code	In UCR/NIBRS?	In NCVS?
11.1 Violations of military law	✗	✗
11.2 Violations of tribal law	✗	✗
11.3 Torture	✗	✗
11.4 Piracy	✗	✗
11.5 Genocide	✗	✗
11.6 War crimes	✗	✗

Category 11 of the classification is comprised of several special-jurisdictional offenses that should—generally and ideally—be extremely rare in practice. Subcategories such as piracy and genocide are primarily included in the classification to retain compatibility with the International Classification of Crime for Statistical Purposes (ICCS) on which it is patterned, and also because they are named and defined offenses in the federal criminal code.

The definitions in our proposed classification for 11.1 violations of military law and 11.2 violations of tribal law are included fit the bill of capturing offenses "not elsewhere classified" in the categories. Both U.S. military law and American Indian tribal law largely parallel federal and state criminal codes in defining offenses—and so the intent would be for military and tribal authorities to report those offenses for national compilation and analysis—but they also have the latitude to define infractions that are unique to their populations and jurisdictions. Relevant to implementation of more effective crime reporting, both broad categories also raise important concerns related to access to offense information, regardless of whether the offense is or is not elsewhere classified in our taxonomy. Taking them in turn, to provide slightly more detail:

- As profiled in Section 2.3.1 of Report 1, the Uniform Code of Military Justice (UCMJ; 10 USC §§801–946) governs the conduct of U.S. military personnel, including personnel working at any U.S. military installation. The UCMJ's "punitive articles" directly parallel other crimes listed in this classification, the key difference being the manner in which the offenses are adjudicated (e.g., in courts-martial) and the penalties that may apply (e.g., including termination of service). Yet others are unique to the UCMJ, as described more completely in Report 1's Box 2.5—such offenses as absence from duty, unlawful enlistment into the service, and refusal to comply with orders. These offenses are, consistent with the general behavioral definitions we outlined in Report 1, crimes that warrant inclusion in the overall picture of crime in the United States.

  The U.S. Department of Defense maintains its own incident-based record system—a data resource covering all of the offenses in the UCMJ and handled by the Defense Department's police and justice authorities. Accordingly, the Defense Incident-Based Reporting System (DIBRS) is uniquely capable of both providing measures for category 11.1 but also including crimes at military installations or including military personnel in the statistics for other offenses as well. As Section 2.3.1 of Report 1, presentations to our panel, and a report from the U.S. Department of Defense, Inspector General (2014) indicate, linkage between DIBRS and NIBRS has been pursued for several years but remains elusive, largely hampered by a technical crosswalk between the two systems, ensuring that incidents are not counted (or double-counted) in surrounding cities or states due to clashing geographic descriptors.
- Law enforcement and justice system processing in Indian Country—the term used to describe American Indian tribal lands and the self-governing, sovereign communities within them—is complex. As self-governing entities, tribes commonly maintain their own police forces, courts, and other authorities, and also typically have their own penal or criminal code defining offenses. The actual handling of criminal proceedings on tribal lands or involving tribal members depends on the nature of the offense, the location of the offense, and the tribal membership (or lack thereof) of both victim and offender. Pursuant to the U.S. Supreme Court's ruling in *Oliphant v. Suquamish Indian Tribe* (435 U.S. 191 [1978]), tribal police and courts lack jurisdiction over non-Indians even if they commit

offenses on tribal lands, and so they would not be empowered to make arrests (or, presumably, record or report the offenses). Conventionally, post-*Oliphant*, arrests of non-Indian perpetrators offending against non-Indian victims on tribal lands would have to be made by county or state authorities; arrests of non-Indians offending against tribal members would have to be made by U.S. federal authorities.[14]

In terms of data collection related to 11.2 violations of tribal law, the Tribal Law and Order Act of 2010 (P.L. 111-211) vests lead authority for the collection of statistics on crime in Indian Country in BJS; Section 251(b) of the act amends the numerous, listed duties of BJS (42 USC § 3732) to include reference to tribal governments as well as federal, state, and local. More specifically, the act adds a general clause to BJS's duties:

> Consultation with Indian Tribes.—The Director [(of BJS)], acting jointly with the Assistant Secretary for Indian Affairs (acting through the Office of Justice Services) and the Director of the Federal Bureau of Investigation, shall work with Indian tribes and tribal law enforcement agencies to establish and implement such tribal data collection systems as the Director determines to be necessary to achieve the purposes of this section.

C.2 CORRESPONDENCE OF SUGGESTED ATTRIBUTES TO CURRENTLY COLLECTED DATA ELEMENTS

The listings below summarize the way in which the attributes suggested in Section 5.2.2 of Report 1 correspond to currently collected data elements in NIBRS. If the element is defined in NIBRS, its number in the NIBRS database structure is listed in square brackets following the name (e.g., [N4] for NIBRS Data Element 4). Unless otherwise cited, references to NIBRS attributes are derived from Federal Bureau of Investigation (2015).

The NIBRS list of attributes/elements is supplemented by elements defined in one or more state-specific incident-based reporting systems. The systems mentioned in this way are from six states and were those for which specification manuals were readily available; all six state systems generally include the full set of NIBRS elements. The six states are:

- Michigan (MI; Michigan Incident Crime Reporting Unit, 2016),
- Montana (MT; Montana Board of Crime Control, Statistical Analysis Center, 2015),

[14]That said, some state laws allow for cross-deputization of tribal law enforcement officers, permitting them (and the tribal law enforcement agency) to arrest and process non-Indians committing offenses on tribal lands, subject to county or state law and prior to transfer to county or state authorities. This is in the same manner that local law enforcement agencies (e.g., in neighboring suburbs) may formally agree to share arrest privileges within their separate geographical jurisdictions.

- New York (NY; New York Office of Justice Research and Performance, 2016),[15]
- Tennessee (TN; Tennessee Bureau of Investigation, 2013),
- Texas (TX; Law Enforcement Support Division, Texas Department of Public Safety, 2015), and
- Virginia (VA; Criminal Justice Information Services Division, Virginia State Police, 2011)

In addition, in selected instances, we refer to the codes and information collected by the quasi-incident based Supplementary Homicide Report (SHR) component of the UCR Program—the table of additional incident-level data items requested for every murder reported in an agency's monthly Return A (the Summary Reporting System's base form).

C.2.1 Incident-Level Attributes

Points of Agreement Between Recommended Classification and NIBRS

- Incident date and time [N3]: We suggest collection of incident time at least to precision of the nearest hour of the day, as in NIBRS. Incident time is recorded in the NY, TN, TX, and VA systems (to the nearest hour in VA). NIBRS Data Element 3 records a single date/time but can also include an "R" flag to indicate that the date/time is the time a report was made, if the incident time is unknown.
- Incident location type [N9]: In Section 5.2.2 of Report 1, we suggest a rough, starting minimum set of location type codes—a rearrangement and regrouping of the 58 location types coded in NIBRS—as follows:
  - Residential location
  - Store or retail
  - Financial institution
  - Commercial establishment (e.g., commercial or office building, hotel, or other lodging)
  - Entertainment venue
  - School grounds and academic buildings
  - Civic or justice system establishment
  - Transportation or related facility
  - Outdoors

[15]The New York State Incident-Based Reporting (NYSIBR) system was, as of mid-2016, in the midst of improvements to ensure full compliance with NIBRS. This improvement project involves adding variables to indicate cargo theft involvement and criminal activity/gang involvement and substantially revising 8 other data elements (including bias crime type, weapon/force used, and the race of victim, offender, and arrestee). See http://www.criminaljustice.ny.gov/crimnet/ojsa/crimereporting/ibr_refman/Notice-of-Change-NYSIBR-03-18-2016.pdf.

- Other physical location (e.g., abandoned/condemned structure, tribal lands)
- Multiple physical locations
- Not location-based (e.g., cybercrime)

- Number of victims, all offenses in incident: Not an explicit data element in NIBRS, but derivable from the number of victim identifiers on the NIBRS data record
- Number of offenders, all offenses in incident: Similarly to the number of victims, not an explicit data element in NIBRS but one that is derivable

Attributes in Recommended Classification but Not in NIBRS

- Incident geographic location: In Report 1, we suggest collection, "ideally, [of] an appropriately anonymized latitude/longitude pair" or—barring that—"a geocode to some small-area geography such as census block or tract."
  - TN: Physical street address where the incident occurred or latitude/longitude coordinates if known
  - NY: City/town/village where the incident be coded, with an identifier for the law enforcement agency's station, division, or precinct being optional
  - MI: County and city/township of occurrence, with an option "geo-code" slot that could be used to "identify station, division, post, team, precinct, etc." (Michigan Incident Crime Reporting Unit, 2016:14)
- Extraterritoriality: Yes/No indicator of whether incident includes offenses that may have been committed, executed, or initiated in another country or state (though federal or state law may hold that such offense may be considered as though committed within local jurisdiction)
- Multiple jurisdiction: Yes/No indicator as to whether incident includes offenses that could be legitimately counted or reported in more than one law enforcement agency's operational jurisdiction

Data Elements in NIBRS but Not in Recommended Classification

- Originating Agency Identifier (ORI) [N1]: Identifier code for the reporting agency
- Incident number [N2]: Unique identifier number assigned by reporting agency to each incident in submission
- Offender suspected of using alcohol, drug/narcotics, or computer equipment [N8]: Incident-level flag for whether any of the offenders appeared to have used alcohol/drugs or—for whatever reason fusing the very different concepts—whether computers were involved. The NY system simplifies this variable to "offender used computer"

- Cargo theft [N2A]: yes/no indicator of whether incident involves the theft of any cargo "that constitutes, in whole or in part, a commercial shipment of freight [that is] moving in commerce" or "in the supply chain"
- Cleared exceptionally [N4]: indicator of whether the agency cleared the incident by "exceptional means," such as declined prosecution, change in custody to another jurisdiction, or death of the offender
- Date of exceptional clearance [N5]

Elements Collected in Other Incident-Based Reporting Systems

- Date and time of filing of incident report [NY]: As indicated above, NIBRS allows a single date/time to be recorded, with a flag to indicate if it is the reporting time rather than the incident occurrence time. The NY system, at least, resolves ambiguity by entering two date/time combinations. In NY, the date/time of filing may also be construed as a "delimiter for the end of the incident," for an incident occurring over a span of days (New York Office of Justice Research and Performance, 2016:4)
- Incident case status [NY]: superset of NIBRS's variables for exceptional clearance, allowing for coding of (normal) clearance by arrest or other statuses (i.e., warrant issued, investigation pending, closed)
- Incident complete indicator [MI]: yes/no, where yes connotes "incident is complete, no further data to immediately follow" and no connotes "incident is not complete, supplemental data to immediately follow" (Michigan Incident Crime Reporting Unit, 2016:101)
- Identity theft-related [TN]: yes/no flag, applicable when an incident involves certain offenses including fraud and larceny/theft, as to whether an element of identity theft was involved in the offense
- *Incident-level elements specific to family or domestic violence*
  - Family violence indicator [TX]: yes/no flag, coded (or derived) when the incident includes assault, homicide, kidnapping/abduction, robbery, or sex offenses and at least one victim-offender relationship (as entered under victim attributes) is a familial or intimate partner relationship
  - Domestic violence indicator [TN]: yes/no, defined similarly to Texas' incident-level family violence indicator; Montana's system also includes a yes/no "domestic abuse related" per-offense variable; Michigan's yes/no domestic violence indicator is associated with each victim
  - Previous domestic violence [TN]: yes/blank indicator to be completed only if the offense-level domestic violence flag is "yes," if the arrestee has been arrested in previous domestic violence incidents involving the current partner/victim or anyone else
  - Warrant signed by [TN]: for domestic violence arrests only, indicator as to whether the victim, a law enforcement officer, or both signed the warrant

- Was the victim transported to a safe place/away from the scene? [TN]: yes/no
- Did incident involve a violation of an order of protection? [TN]: yes/no

- *Incident-level elements specific to officer-involved shootings*
  - Officer-involved shooting flag [TN]: yes/no
  - Outside investigation? [TN]: yes/no, whether the incident/shooting was investigated by an outside agency
  - Officer sequence number [TN]: sequential number assigned to each officer who fired a weapon during the incident
  - Officer's weapon type [TN]: firearm (type unknown), handgun, rifle, shotgun, other firearm (machine gun, bazooka, etc.)
  - Number of shots fired by the officer [TN]
  - Distance between officer and suspect, in feet [TN]
  - Officer's type of assignment [TN]: uniform, plain clothes, special assignment
  - Multiple agencies involved [TN]: yes/no, whether officers who fired weapons during the incident were from different law enforcement agencies
  - Officer's number of years on force [TN]: years of employment with current agency
  - Incident level [TN]: indication of whether incident was a felony or a misdemeanor
  - Law Enforcement Officers Killed or Assaulted (LEOKA) involvement [TN]: yes/no flag, yes if victim was law enforcement officer

C.2.2 Per-Offense Attributes, General

Points of Agreement Between Recommended Classification and NIBRS

- UCR Offense code [N6]: Entry of an offense code is not explicitly rendered in our recommended attribute list—treating codes and attributes as coequal parts of overall classification—but is certainly tacit. New York's system requires coding from a state-issue coded law file (denoting the section of law and the degree of the offense), which also subsumes the attempted/completed flag mentioned below. A separate variable is required for larceny-type offenses, to match NIBRS's categories, because these offenses cannot be distinguished by the law-file codes.
- Offense attempted/completed [N7]: This is a case of partial agreement, in that our recommended attribute lists adds responses for threatened, not applicable, and not known
- Apparent/suspected bias motivation [N8A]: NIBRS permits entry of up to five motivations per offense, with different categories for race/ethnicity, religion, sexual orientation, gender, gender identity, or disability. At

least for the outset, our recommended attribute list suggests focusing on yes/no indicators for these higher level bias-motivation categories, as well as adding four other entries: professional affiliation (e.g., justice system/law enforcement), other bias, no apparent bias motivation, or not known

- Type of weapon or force involved, for attack-type offenses [N13]: Combining the codes that are deemed valid for NIBRS Data Element 13 with codes derived from SHR data yields the set:
  - 11 Firearm (type not stated; code can include "A" suffix to indicate automatic weapon)
  - 12 Handgun—pistol, revolver, etc. (code can include "A" suffix to indicate automatic weapon)
  - 13 Rifle (code can include "A" suffix to indicate automatic weapon)
  - 14 Shotgun (code can include "A" suffix to indicate automatic weapon)
  - 15 Other firearm (code can include "A" suffix to indicate automatic weapon)
  - 20 Knife or cutting instrument (knives, razors, hatchets, axes, cleavers, scissors, glass, broken bottles, ice picks, etc.)
  - 30 Blunt object (baseball bats, butt of handgun, clubs, bricks, jack handles, tire irons, bottles, etc.)
  - 35 Motor vehicle (valid value in NIBRS but not SHR)
  - 40 Personal weapons (hands, fist, feet, arms, teeth, etc.)
  - 50 Poison (in SHR coding, does not include gas; death by gas would be considered 85 asphyxiation)
  - 55 Pushed or thrown out window (included as SHR code, but not NIBRS)
  - 60 Explosives
  - 65 Fire/incendiary device
  - 70 Drugs/narcotics/sleeping pills
  - 75 Drowning (included as SHR code, but not NIBRS)
  - 80 Strangulation—hanging (included as SHR code, but not NIBRS)
  - 85 Asphyxiation—includes death by gas
  - 90 Other (SHR coding includes unknown weapon; NIBRS examples include BB guns, pellet guns, tazers, pepper spray, stun guns, etc.)
  - 95 Unknown (valid value in NIBRS but not SHR)
  - 99 None (valid value in NIBRS but not SHR)

  Our recommended classification suggests a more streamlined list as a minimum:
  - No weapon or force involved
  - Weapon or force involved

  - Attack with firearm:[16] Handgun/pistol; rifle/long gun; other/unknown
  - Attack with non-firearm weapon: Knife, cutting instrument, or sharp object; blunt object; motor vehicle; poison; fire or incendiary device; explosives; drugs or narcotics; other external weapon/force
  - Attack without external weapon: Bodily attacks (hands/fists, feet, etc.); asphyxiation or strangulation; drowning or submersion; pushing into harm's way (e.g., from high place)
 - Unknown
- *Type of property involved* (stolen, damaged, etc.) in offense [N15]: Ideally, a simplification of existing 80 or so codes, while adding "intangibles" such as intellectual property and personally identifiable information
- Type of drug/psychoactive substance involved in offense [N20] and quantity [N21]: NIBRS includes a third element, type of drug measurement [N22], to indicate the unit of measurement; Tennessee adds specific qualifier of this being the "suspected" drug involved. Our classification recommends (but does not make explicit) a variant on the roughly 20 drug types/classes currently coded in NIBRS; in the current NIBRS, up to 3 drug types can be specified and an indicator raised to indicate involvement of more than 3 drug types.

Attributes in Recommended Classification but Not in NIBRS

- Victim/offender situation: Summary indicator of whether offense involves single or multiple victims, and single, multiple, or an unknown number of offenders
- Group support/motivation: Indicator of whether offense is organized (gang-related, organized-crime related, terrorism-related, unknown organized involvement), unorganized, not applicable, not known
- Cybercrime-related: Yes/no flag indicates whether the use of computer data or computer systems was an integral part of the modus operandi of the offense. This variable disentangles elements in the curious NIBRS Data Element 8, "offender suspected of using" alcohol, drug/narcotics, or computer equipment and so might be considered a partial match.

Data Elements in NIBRS but Not in Recommended Classification

- Type criminal activity/gang information [N12]: Complicated element that applies only to 8 NIBRS categories for which up to 3 types of activity (e.g., buying/receiving, exploiting children, operating/promoting/assisting), or only to 11 additional NIBRS offenses for which up to 2 gang

[16]Virginia's incident-based system includes an indicator as to whether the firearm used is an automatic weapon.

information codes can be indicated (juvenile gang, other/adult gang, none/unknown). Our recommended attribute list includes a streamlined group support/motivation attribute, described above.
- *Elements applicable only to burglary offenses:*
  - Number of premises entered [N10]: used only when "Hotel Rule" applies
  - Method of entry [N11]: force or no force
- Type of property loss [N14]: e.g., burned, destroyed/damaged/vandalized, seized. This information is assumed to be conveyed through the offense code.
- Number of stolen and recovered motor vehicles [N18, N19]
- Value of property [N16]: total dollar estimate, with default value of $1 if unknown (Federal Bureau of Investigation, 2015:105); exception for entry generally made for drugs/narcotics seized in a drug offense (e.g., in Tennessee)
- Date property recovered [N17]

Elements Collected in Other Incident-Based Reporting Systems

- Home invasion [TN]: yes/no flag applicable to burglary and robbery offenses only, defining "home invasion" as occurring "when one or more subjects enter an occupied dwelling with the intent to commit an offense therein, and in the course of their entry or presence in the dwelling, commit one or more violent offenses (e.g., rape, assault, etc.) against the occupants of the dwelling" Tennessee Bureau of Investigation (2013:13)
- Type of security used at scene of offense [VA]: up to 2 entries, from alarm/audio, alarm/silent, bars/grate, camera, dog, deadbolt, locked, unlocked, exterior lights, interior lights, fence, guard, Neighborhood Watch, other, none
- Offender's method of entry/exit to building/structure [VA]: up to 2 codes for entry and 2 for exit, from: front, rear, side, attic, vent/air conditioner, window, door, patio/sliding door, balcony/fire escape, attached garage, wall, vehicle, floor, roof/skylight, hidden within, other, unknown
- How offender left scene [VA]: auto, truck, van, motorcycle, bicycle, foot, moped, other, unknown
- Involvement of other factors [MT]: in addition to a yes/no flag for "domestic abuse related," Montana's system requires yes/no answers to the involvement of three other factors in each offense: gang related, gambling related, and mental health related. The Montana system also includes an indicator as to whether the offense was investigated by or referred to the state's Drug Task Force.

- *Other means of valuation of property involved in offense:*
  - Property value over $500 [TN]: may be hard-coded yes/no, or derived from entry for total value of property
  - Recovered value [TN]: value of recovered stolen property, which can be less than or equal to the estimated value lost
- *Other means of quantifying information about drugs/controlled substances involved in offenses:*
  - Type of drug/narcotic seized in a drug case [TX]
  - Estimated quantity of drug/narcotic, and type of measurement used in that quantity, seized in a drug case [TX]
  - Drug origin type [TN]: clandestine lab, diverted, illegally imported (from any site outside the immediate area of the drug seizure, not necessarily from outside the U.S.), indoor grow, outdoor grow
  - Nature of marijuana seized, and number of such sites [TN, TX]: Texas' system includes two variables indicating whether the marijuana was found in gardens, cultivated fields, wild fields, or greenhouses, and the number of such cultivation sites seized; Tennessee's asks "where the majority of the plants were found, the number of plots, and, if known, the longitude and latitude of the plots," as well as a basic flag as to whether the plants were being grown outdoors, indoors, or both (Tennessee Bureau of Investigation, 2013:34)
  - Number of clandestine labs seized [TX]
  - Type of drug manufactured in seized clandestine labs [TX]: methamphetamine, amphetamines, phenylacetone (P2P), phencyclidine (PCP), crack cocaine, other
  - Methamphetamine precursor [TN]: for use only when the drug type is methamphetamine and the drug origin type is clandestine lab; possible values are single entity tablet/capsule; combination tablet/capsule; gelcap; liquid; unknown/not present
  - Quantity of precursor chemical seized, and unit of measurement used in that quantity [TX]
  - Source of drug data [NY]: indicator of whether the information about drugs seized during an incident is an estimate (by the officer or agency) or a laboratory analysis; values of "not applicable" or unknown are also permitted

C.2.3 Per-Offense Attributes, Victim/Offender Relationship

Relationship between each combination of victim and offender, for all the offenses coded within a particular incident, is certainly one of the most useful analytical variables of the resulting data and is essential to eventual reanalysis/reclassification of offense classes such as domestic or intimate partner violence. Accordingly, it is an attribute that is central to our proposed classification as well as current data collection.

The alphabetic codes listed below are those used in NIBRS Data Element 35, relationship of victim to offender. The UCR's Supplementary Homicide Report (SHR) component also collects victim-offender relationship for each specific occurrence of murder, excluding some of the NIBRS codes and generally splitting relationship categories into gender-specific subcategories, as is also shown below. The SHR alphabetic codes are from Federal Bureau of Investigation (2012), which is the codebook for the most recent Inter-University Consortium for Political and Social Research (ICPSR) archive of SHR data. However, it should be noted that the actual SHR instrument (the FBI's Form 1-704) asks for free-text response in a table column headed "Relationship of Victim to Offender (Husband, Wife, Son, Father, Acquaintance, Neighbor, Stranger, etc.)," and the alphabetic codes are derived later in processing.

Within Family
SE Spouse (SHR: HU husband and WI wife)
CS Common-Law Spouse (SHR: CH common-law husband and CW common-law wife)
PA Parent (SHR: FA father and MO mother)
SB Sibling (SHR: BR brother and SI sister)
CH Child (SHR: SO son and DA daughter)
GP Grandparent (not defined in SHR)
GC Grandchild (not defined in SHR)
IL In-law
SP Stepparent (SHR: SF stepfather and SM stepmother)
SC Stepchild (SHR: SS stepson and SD stepdaughter)
SS Stepsibling (not defined in SHR)
OF Other Family Member

Outside Family But Known to Victim
AQ Acquaintance
FR Friend
NE Neighbor
BE Baby/Child in the care of a Babysitter (not defined in SHR)
BG Boyfriend/Girlfriend (SHR: BF boyfriend and GF girlfriend)
CF Child of Boyfriend or Girlfriend (not defined in SHR)
HR Homosexual Relationship (SHR uses code HO)
XS Ex-Spouse (SHR: XH ex-husband and XW ex-wife)
EE Employee
ER Employer
OK Otherwise Known

Not Known By Victim
RU Relationship Unknown (SHR: UN relationship not determined)
ST Stranger

Other
VO Victim Was Offender (not defined in SHR; meant to be invoked in such cases as "domestic disputes where both husband and wife are charged with assault, double murders

(two people kill each other), or barroom brawls where many participants are arrested")

By comparison, our recommended classification suggests a slightly simplified, minimum set of categories. Specifically, our victim/offender relationship attribute codes whether the suspected perpetrator is, to the victim:
- Stranger
- Relative: Parent; child; sibling; grandparent; grandchild; other relative
- Known to the victim: Current spouse or intimate partner; former spouse or intimate partner; colleague (e.g., at workplace); friend; other known person/acquaintance
- Unknown

C.2.4 Victim Attributes

Points of Agreement Between Recommended Classification and NIBRS

- Type of victim [N25]: Our proposed classification uses subcategories personal/individual, business, financial institution, government, religious organization, law enforcement officer, other, and unknown
- Age of victim [N26]: Implemented in NIBRS as a two-character alphanumeric code—01–98 corresponding to years of age, NN under 24 hours, NB 1–6 days old, BB 7–364 days old, 99 over 98 years old, and 00 unknown
- Sex of victim [N27]: Labeled "gender of victim" in proposed classification
- Race and ethnicity (Hispanic/non-Hispanic origin) of victim [N28–29]: Proposed classification calls for categories to defined in accordance with "race/ethnicity categories used by the U.S. Census Bureau and/or required by U.S. Office of Management and Budget
- Resident status of victim with respect to operational jurisdiction of reporting agency [N30]: resident, nonresident, unknown

Attributes in Recommended Classification but Not in NIBRS

- Citizenship of victim: Categorized in recommended classification as U.S. (including dual with U.S.), foreign, refugee/no citizenship, and not known

Data Elements in NIBRS but Not in Recommended Classification

- Victim sequence number [N23]: Identifier for each victim involved in incident; not explicitly rendered in our recommended classification, but tacit
- Victim connected to UCR offense code [N24]: Up to 10 offense codes perpetrated against each specific victim

- Type of injury [N33]: for 11 UCR crime types, one or more codes for the level of physical injury: none, apparent broken bones, possible internal injury, severe laceration, apparent minor injury, other major injury, loss of teeth, unconsciousness. New York's system adds a second "level of injury" variable, with values: dead, appeared to be seriously injured, appeared to be physically injured but not seriously, not injured, not applicable, not reported, unknown. Extent of injury is integral to the revised definitions of assault in our recommended classification, hence exclusion as a separate attribute.
- Offender number to be related [N34]: identifier of up to 10 offenders listed in data record to be associated with each victim
- Aggravated assault/homicide circumstances [N31]: up to 2 (of 10) circumstances per victim, including argument, "lovers' quarrel," and mercy killing; 1 (of 5) circumstances for negligent manslaughter, including gun-cleaning accident or hunting accident; or 1 of 2 circumstances for justifiable homicide (criminal killed by private citizen or killed by officer). For homicide, the Supplementary Homicide Report (SHR) that is required by the UCR program even of Summary-responding jurisdictions, text responses to a "circumstance of homicide" variable are coded into a different set of codes, as described in Box C.1.
- Additional justifiable homicide circumstances [N32]: e.g., criminal attacked police officer and was killed by another police officer or criminal attempted flight from a crime
- Elements specific to law enforcement officers assaulted or killed in the line of duty:
  - Type of officer activity/circumstance [N25A]: for law enforcement officers assaulted or killed in line of duty, code for activity at time of the offense (e.g., responding to disturbance call, handling/transporting/custody of prisoners, ambush/no warning)
  - Officer assignment type [N25B]: e.g., one-officer vehicle (alone), two-officer vehicle, detective or special assignment
  - Officer–ORI of other jurisdiction [N25C]: ORI for officer's home agency, if assaulted or killed in different jurisdiction

Elements Collected in Other Incident-Based Reporting Systems

- Residence status of victim with respect to the county/city where incident took place [MI, NY]: Alternate approaches to the concept of residence status. New York's system includes a variety of living situation codes: resident, commuter, tourist, military, student (either living outside the area or having "temporary domicile in the jurisdiction only because of school attendance"), temporary resident/foreign national, other status, homeless, not applicable, not reported, unknown (New York Office of Justice

Research and Performance, 2016:60). Michigan's victim residence variable is narrower, permitting values of: resident of community/city/town of occurrence; resident of county but not the community of occurrence; resident of state but not the county or community of occurrence; out of state; unknown.

- Attributes specific to victimizations of college students:
  - Is victim a college student? [TN]: yes/no; Tennessee requires that agencies report "information for college students attending colleges/universities located in the same COUNTY as the agency (even if the college is in a different CITY)" (Tennessee Bureau of Investigation, 2013:49)
  - Name of college or university [TN]: if victim is a college student
  - Did the offense occur on campus? [TN]: yes/no
  - Victim type code [TN]: student, faculty/staff, guard/security, other

C.2.5 Offender Attributes

Points of Agreement Between Recommended Classification and NIBRS

- Age of offender [N37]
- Sex of offender [N38]: As gender of offender
- Race and ethnicity (Hispanic/non-Hispanic) of offender [N39 and 39A]

Attributes in Recommended Classification but Not in NIBRS

- Type of offender: Same categorization as type of victim, and important to retain as standalone attribute to emphasize classification's coverage of nonperson actors.
- Residence status of offender with respect to operational jurisdiction of reporting agency
- Citizenship status of offender

Data Elements in NIBRS but Not in Recommended Classification

- Offender sequence number [N36]: identifier for each offender involved in incident
- *Elements specific to arrestee, as distinct from suspected offender:*
  - Arrestee sequence number [N40]: identifier for each arrestee involved in incident
  - Arrest transaction number [N41]: unique identifier assigned by agency to arrest report; Tennessee's system includes provision for entering the state-assigned control number from an arrestee's fingerprint card
  - Arrest date [N42]

- Type of arrest [N43]: on-view arrest (apprehension without warrant/previous incident report), taken into custody (apprehension based on warrant/previous incident report), or summoned/cited (not taken into custody)
- Multiple arrestee segments indicator [N44]: if an arrest is related to multiple incidents (different data records), then this variable is intended to be coded C for "count arrestee" in one incident and M ("multiple") in any other incidents, so that arrests are not multiply counted
- UCR arrest offense code [N45]: single, most serious offense for which offender was arrested
- Arrestee was armed with [N46]: code for type of weaponry, if any, arrestee was carrying at time of arrest
- Age of arrestee [N47]
- Sex of arrestee [N48]
- Race and ethnicity of arrestee [N49–50]
- Resident status of arrestee with respect to operational jurisdiction of reporting agency [N51]: resident, nonresident, unknown
- Disposition of arrestee under 18 [N52]: handled within department (e.g., released to parents or released with warning) or referred to other authorities (e.g., referred to juvenile court)
- Arrest clearance indicator (yes/no) and specific UCR offense code cleared by arrest

Elements Collected in Other Incident-Based Reporting Systems

- Offender condition [NY]: meant to characterize the "*alleged* condition" of each offender, with values as follows: appears to be impaired with drugs, appears to be impaired with alcohol, appears to be mentally disordered, appears to be injured/ill, apparently normal, not reported, unknown (New York Office of Justice Research and Performance, 2016:51)

– D –

Remaining Methodology and Implementation Issues for Modern Crime Statistics

The themes raised in this appendix are vitally important concepts, based on our interactions with crime data users and stakeholders. Included among these themes are what is arguably the single most important and pervasive use of national crime statistics and what we suggest might be the single most important benefit of national crime statistics collection. These are weighty topics that underscore both the difficulty and the importance of modernizing crime statistics, and that warrant a fuller airing than brief mentions in earlier text. We provide details on these issues in this appendix because exploring them at appropriate detail in Chapters 2 and 3 would be unduly disruptive to the overall flow of arguments there.

In Section D.1, we review the historical and current implementation barriers—real and perceived—that hinder the assimilation of data (from state and local data resources or other sources) at the national level, focusing on the development of the National Incident-Based Reporting System (NIBRS). We also describe what has been done to remove those barriers to change in national crime statistics. In Section D.2 we posit that attention to data and information quality should be billed as the primary value added by national-level integration of crime statistics. In Section D.3, we suggest presenting crime information in a manner that allows for (not undercuts) the primary use to which stakeholders put crime statistics—comparisons in crime across jurisdictions and over time. A final major theme in Section D.4 concerns the special problems in characterizing crime involving businesses or organizations as actors in crime, and how improvements in the documentation of crime affecting business might be obtained.

D.1 AVOIDING THE NIBRS DEVELOPMENT PITFALLS

In Section B.6 (Appendix B), we mentioned early research that tried to articulate the barriers—real or imagined—that slowed adoption of the Uniform Crime Reporting (UCR) Program's National Incident-Based Reporting System (NIBRS) after its introduction in the late 1980s, and we briefly described some of them in Section 2.2.3 (Chapter 2). But these issues demand fuller commentary for two basic reasons. First, the lessons learned from NIBRS's development are applicable more widely—not just to incident-based reporting of crime data from local agencies, but also to other efforts that involve assimilating data from disparate state, local, private, and other sources. Although the language and examples used in this section are almost exclusively UCR/NIBRS-oriented; they serve as an important and telling case study. Many of the same barriers can be expected to occur in the coordination of data collection on other offense types from diverse external sources, such as records of environmental violations or child welfare investigations from state or municipal offices. Many of the same barriers seen in NIBRS development are inevitable in working with administrative and other sources of data to generate measures of the nontraditional offenses in the recommended crime classification. Combining data from multiple state administrative departments, or from the courts, will involve managing patchworks of standards and sets of competing interests, as was the case in NIBRS. The second reason for commentary is that we envision police-report data to continue to be an important, primary source of national crime statistics in our ideal, modernized crime data infrastructure. Accordingly, a pivotal question that must be asked is: How can there be confidence in implementing this long-term vision when 30 years of NIBRS development have resulted in participation levels shy of 40 percent and when some federal agencies have not reported data to the UCR program (as required by federal law)?

For each of the barriers, we briefly suggest mitigation factors—generally, things that are different now that increase the probability of successful adoption of incident-based reporting and innovations in crime statistics as a whole. Two of these potential success factors are sufficiently important, and crosscut many of the historical barriers, that it is best to state them first and then refer to them as needed:

- *The establishment of the National Crime Statistics Exchange (NCS-X) initiative by the Bureau of Justice Statistics (BJS), in cooperation with the*

Federal Bureau of Investigation (FBI): The NCS-X work to stimulate NIBRS-format reporting by a carefully designed sample of jurisdictions, and promoting NIBRS planning and development more generally, is critically important in several ways. Methodologically, it is a partial return to first principles for incident-based police-report crime data—-retrofitting national crime data collection with the start-from-sample approach that had been advocated by the *Blueprint for the Future of the Uniform Crime Reporting Program* (Poggio et al., 1985) at the outset. When finished, the NCS-X additions will finally give NIBRS data the opportunity to demonstrate their merits in that they will be generalizable to the nation as a whole and important subnational groups. The capacity of detailed incident-specific data to examine the dynamics of crime have heretofore only been available on the national basis for homicide (through the UCR's Supplementary Homicide Report [SHR] collection), and the case for active NIBRS expansion has been weakened without it. The NCS-X work has also been important in level-setting and taking stock of the current information management systems capacity among many state and local law enforcement agencies. But most fundamentally it represents the hint, and hopefully the start, of concerted national-level engagement in the compilation of national crime statistics—an active stance toward the production of crime statistics rather than a more passive, aggregation stance, and one consistent with our call for strong coordination and governance in Chapter 3.

- *The FBI's stated decision to retire UCR Summary format reporting in favor of NIBRS by 2021:* In our discussions in 2014, before the decision began to take shape, the panel heard a range of viewpoints on how major changes in police-report data collection *might* be made, and might be made most effectively. Solid arguments could be made for or against either a top-down or a bottom-up approach. For instance, major changes being made by fiat—from on high, as when new crimes (and new forms) have been added to UCR collection in response to legislative mandate—are prone to meet resistance among local law enforcement respondents, who might see the unfunded mandate placed upon them swell larger. But, on the other hand, the bottom-up approach—embodied, in part, by the elaborate FBI Criminal Justice Information Services (CJIS) Advisory Policy Board (APB) approach—is properly grounded in recognition of the voluntary nature of data transfer to national collections. Yet, it can be slow and cumbersome and generally precludes the implementation of quick, major change to adapt to emerging problems. Accordingly, the FBI's decision to retire a system that has held sway for nearly a century is remarkable in its own right. The manner in which it was made is also important, combining elements of both approaches—made a priority by FBI leadership, but predicated on increased recognition among the whole stakeholder base of the inadequacy of UCR Summary counts going forward, done with the endorsement of major law enforcement support organizations, and coursing rapidly through the APB process. Publicly setting an ambitious deadline has associated risks, but—for purposes of overcoming resistance to change—positioning "NIBRS as the only choice" for national reporting is likely the most effective strategy.

Many of the barriers to more rapid NIBRS adoption can be most succinctly summarized as local agencies being—for a variety of reasons—simply unwilling or unable to make the switch. Decomposing the unwilling side first, the most commonly voiced barrier to NIBRS implementation, by the local law enforcement executives who would be responsible for it, is probably the *apprehension about the appearance of crime rate "spikes" in their jurisdictions due to the counting of multiple offenses* in crime incidents: that is, through elimination of the Hierarchy Rule by which only a single offense is recorded. This concern is grounded in the undeniable fact that changing the yardstick of measurement will inevitably have some effect on the measure. The crime numbers will indeed be *different* for some offenses, with increases and offsets to be examined and reconciled. But the testimony we heard from incident-based reporting adopters in our meetings—coupled with the experience of changing crime statistics in other countries as well as the recent U.S. experience in revising the UCR definition of rape earlier this decade—convinces us of a few key, related points. First, our strong impression from early adopters is that real spikes—major, substantive increases in particular offense rates—will likely be closer to the exception than the rule for local jurisdictions. Where they occur, they may speak usefully to and be explained by other data collection issues, such as past misunderstanding or misapplication of relevant definitions. In any event, this is a barrier for which the logical (and effective, from past experience) remedy is simple communication. Best practices for the first several releases of new-format data can and should be developed from previous and early NIBRS adopters, including the importance of building awareness of the forthcoming change among the public, the media, and state and local government officials. This is a topic in which the NCS-X experience—a fairly large and diverse cohort of state and local jurisdictions going through the change simultaneously—will also be helpful, with more localities having managed any short-term spikes in crime rates to contribute to the best practices for conversion.

If the fear of spikes is the most vocalized barrier related to general unwillingness to change, then perhaps the most deeply felt barrier might be *the lack of any clear incentive to change—in the sense that direct benefit to a local agency's daily operations of this change is not clear.* Put more simply, the completely understandable question that a local law enforcement agency would raise—laboring under what is effectively an unfunded mandate and possibly contributing to the UCR Summary solely to meet requirements under state law—is, "what's in it for me?" NIBRS benefits could be always be described elegantly and compellingly—see Box B.3 for the benefits of participation as outlined in 1992—but at a high, abstract level, and not showing how the data supplied to the national collection would be returned in a way that could inform local policies and operations. The way in which NIBRS was rolled out—en masse for contribution by any agency, rather than emphasizing first adoption by a sample as recommended by the 1985 *Blueprint*—had consequences in this regard. The lack of national representativeness in the take-up meant that the compiled NIBRS data were neither published nor analyzed to the full extent possible, leaving the process without any visible feedback loop to the early-adopting local participants. They had no sense that any enhancement or value had been added to their contributed data. States that have their own incident-based reporting systems were able to make use of those data—and generate interesting and useful findings—but not necessarily in a way that would convince other states or localities to see potential gains in efficiency or effectiveness from switching standards.

Related to this barrier, we will argue in Section D.2 for positioning attention to data quality and integrity as a critical part of the value-added benefits by contribution to national crime statistics. This, too, is an area in which the NCS-X sample—and, finally, analyses making use of real data—will help greatly. But, more fundamentally, what has changed in more recent years that makes this barrier more superable than in the past is increased appreciation of data analytics by law enforcement practitioners. In some cases, this is borne out of experience with COMPSTAT or related programs—COMPSTAT being the practice adopted by New York City and other cities of using locally available, point-specific crime data to hold precinct supervisors and other officials accountable for changes in crime in their areas. In others, this comes from the intelligence gained through the mapping of crime over time and, in more well-resourced local departments, the establishment of crime analysis units to carry out the work. Still other departments have adopted the practice of publishing fairly detailed crime information on their websites, on a close to real-time basis, or working with online vendors to host such information resources. In short, our sense from interaction with law enforcement practitioners is that there is a greater recognition of crime data as a basic tool of accountability, internally within departments and externally to the public they serve. In this climate, then, we think that analyses coming from NCS-X NIBRS data will make it easier to make the case to law enforcement executives that incident-based reporting to a national program will accomplish the following:

- Support evidence-based policing and policymaking, by facilitating comparison of crime problems (and eventually the interventions to address them) across jurisdictions;

- Permit benchmarking, or comparison in proper context, to "communities like us" on factors other than raw population size, overcoming a critical limitation of aggregate-count-only UCR Summary data;
- Help in public policy formulation, particularly with respect to the "new" crime areas, in which answers to previously unanswered questions could spur new policies and practices; and
- Improve local law enforcement's awareness and intelligence of their own areas, including by permitting analyses of offenses affecting detailed victim populations in other areas (or nationally) that may be useful in developing interventions in or programs for those population subgroups in the local area. (Through more research on NIBRS data, the broader public may also obtain benefits of added intelligence about crime—for instance, in the form of more useful demographic/geographic offense rates or the estimation of probabilities of victimization for at-risk populations.)

Agencies that have had the wherewithal to devote in-house data analyst time to their own data resources—in some cases, either directly publishing crime statistics on the Internet or enabling the open, public mapping of crime events—might be particularly well poised to exploit the benefits of comparison/contrast with peer agencies based on nationally compiled data. Sharing of those results and acquired expertise would further aid the cause of broader implementation.

A final barrier of local agencies' unwillingness to adopt to a new system is a pure problem of political will, and again is partly a consequence of how NIBRS began on an open-to-all basis rather than targeting initial effort. Many of the same departments that the *Blueprint* authors suggested be emphasized first still remain NIBRS holdouts. *Absent participation by large, major-city departments, smaller jurisdictions might resist change to NIBRS for lack of demonstration/example to follow.* Uniformly, the law enforcement agencies in the nation's largest cities refrained from making the switch, even in states otherwise poised to make the conversion. Whatever the reason (and they varied), the downstream effect on wider participation was undoubtedly considerable. This is an ongoing, lingering problem, and one in which both cross-cutting success factors—the NCS-X work that is finally targeting several of the historical big-city "holdouts" and the strengthening of political will that may come from the 2021 UCR Summary sunset deadline—will hopefully help to address. Two other specific points should be made in this regard, the first of which is that—relative to the historical problem of landing large adopters for incident-based reporting—it is difficult to understate the potential importance of California and Texas (as well as Indiana) committing to meet (or beat) the NIBRS-conversion deadline in enacted law (see Section 3.2 and Appendix E). Second, as we allude to later, this stands as a pivotal moment in time when some large cities and state programs will *have* to take on systems modernization initiatives, because their in-house-developed systems have exceeded their practical lifetime and may be untenable to maintain in modern computing environments. Hence, this may be a uniquely appropriate time to influence the shape of replacement systems, ideally in a way that directly facilitates national-reporting extracts.

Turning to barriers related to local agencies being unable to change to NIBRS in the past, a major sticking point was and remains *concern over the (high) costs of NIBRS implementation, both near-term (the start-up costs of implementation or conversion) and long-term (e.g., ongoing maintenance and training costs).* These fundamental worries could apply with equal force to local agencies of different size and with different degrees of extant technical sophistication:

- Just given the numbers involved—of terminals (mobile or otherwise) needed for the capture of police reports, of software licenses needed, and so forth—major technology upgrades are a costly proposition for even large city police departments, much less smaller sized agencies. Those agencies that invested in home-grown, customized information management systems over the years faced extensive costs in overhauling systems—reengineering that is rarely if ever a simple matter of directly porting source code directly to a new computer environment—and could understandably balk, particularly without a compelling end benefit.
- As we suggested in Appendix B, a small law enforcement agency without a sophisticated information management system (and so wholly unprepared for rapid conversion) could fairly view the completion of even the simple UCR Summary forms as a major task of intricate bookkeeping. But the complement to that statement is also true, and partially speaks to cost-based concern in making the switch: Having been conditioned for so long to comply with the intricate UCR Summary calculations, a local agency might find a change to NIBRS potentially overwhelming simply because it tries to fix what isn't (operationally) broken, and would instead upend decades-old administrative routines and work-arounds.
- Generally, the strong impression that we were left with from our interactions with law enforcement practitioners is that the marketplace for law enforcement–specific automated information systems is growing in size but that the number of technical solution providers (i.e., systems vendors) is relatively small,[1] with many of them listing NIBRS-type reporting capability as a selling point. Indeed, it is likely that numerous

[1]In our meetings and discussions, we heard references to numerous vendors or solution providers working in the area of law enforcement–specific information management, comprising both records management and computer-aided dispatch. The solution providers vary in the extent to which they enable mobile-based reporting or use cloud computing capabilities, and to which they explicitly list NIBRS-format output as a key feature. We did not examine the features or

state and local departments that have bought information management solutions are, as a result, equipped with systems that have detailed NIBRS-type reporting built into their systems but the capability is simply not used—possibly because it is an additional-cost licensing add-on. Getting those departments and systems ready for NIBRS submissions is a markedly different proposition than one in which the NIBRS-format reporting would have to be engineered anew in system code.

- Put most colloquially, the opening phase of NIBRS development came with neither carrot nor stick, in terms of suasion and incentive to participate. Substantial seed funding was not available to state or local departments, but neither was reporting in NIBRS format (or the development of plans to do so) linked in any way to fund awards under the Edward Byrne Memorial Justice Assistance Grant (JAG) or Community Oriented Policing Services (COPS) Office grant programs. As discussed in Chapter 3, the underlying reason for not employing a firmer hand—respecting the longstanding and inherently voluntary nature of data contribution to the national reporting program—is eminently reasonable, but the fact remains that the tactic left local agencies with no incentive of any form to adopt.
- As mentioned at the outset, costs *after* the initial conversion/adoption are a major concern. Computers age, systems require update and maintenance, new and continuing staff require training, and so forth—all of which can be daunting to local law enforcement executives under perennial budgetary pressure to curb costs of management relative to reducing the number of officers on the beat.
- Though it is instinctive to be drawn to the problem of systems at the local police level—contemplating systems development within and across thousands of agencies—the challenge of systems upgrade and maintenance at the state level should not be discounted. Changes to UCR/NIBRS content or to reporting requirements imposed by state law necessarily involve additional costs for the state coordinating programs, and is made more complicated because the state UCR coordinating body is typically also the group involved in populating criminal history record databases

capabilities in depth, relying on the interpretations of the law enforcement practitioners involved in our panel's meetings, and we do not claim to have performed an exhaustive inventory. But, as support for assertions such as the law enforcement information management systems vendor space being complex yet still relatively small in size—it is appropriate to mention that solution providers that came up in our meetings include: Motorola (including subsidiary Spillman Technologies, which branches more into crime data analysis and mapping), Integraph, COPLINK (originally a joint venture of the University of Arizona and the Tucson Police Department, now developed by IBM's i2 subsidiary), OSSI/Open Software Solutions (now developed by SunGard Data Systems), TriTech, TylerTech, and New World Solutions. Inclusion in this listing does not constitute endorsement or approval of any sort, and exclusion is likewise not intended as a slight in any fashion.

for the state. As with the systems used by local agencies for records management, there is no direct incentive for the vendors of software used for state information repositories to change their software for free.

The issue of start-up/conversion costs, in particular, is one that is directly addressed by the NCS-X initiative. After years of relative stagnation in NIBRS take-up, the more hopeful recent take-up has been concomitant with the show of national-level commitment to NIBRS that is embodied in the NCS-X work. To the extent that justice system improvement grants will have the resources to continue to provide material assistance to ongoing system maintenance and updating, the NCS-X work is also perhaps the start of a mitigation of the concern about ongoing costs. Another mitigation strategy related to managing overall costs is to focus additional effort on work with, and recommending minimum information system requirements for, the vendors providing hardware and software solutions to both state and local law enforcement agencies. We recognize that there remains heterogeneity in the adoption of information management systems across agencies and from different vendors, as well as the custom, legacy solutions and internal paper reporting that exists in some places. Indeed, our discussions with local agency representatives about the babel of reporting systems and styles that can exist within the same region means that, from the day-to-day investigative standpoint, the only real mechanism for direct information sharing by agencies—still today—is through convening in-person case coordination meetings, simply because systems do not readily "talk" to each other. So there is a great distance to be traveled before national crime statistics collection through automated, direct harvest of state and local records is possible. Reducing the dimensionality of the problem by focusing on the unique individual constraints and demands on information systems vendors is an important step, just as reducing dimensionality by coordination with states is essential for coordinating the inflow of crime data. Work with vendors/solution providers was absent in the early, formative days of NIBRS, but returning to it now might have collateral benefit in shaping the kinds of approaches that might be necessary to bring the smallest and least-resourced law enforcement into the NIBRS and attribute-based reporting fold, such as shared, regional reporting systems and the provisioning of crime reporting via cloud computing.

Another barrier on the "unable" side of the ledger is that—particularly in the late 1980s, but still true to a degree today—*the format and structure of NIBRS records and data is complex, representing a daunting computational task and an inflexible design.* Before relational databases and other computational innovations truly took hold in the broader market, law enforcement agencies could understandably blanch at suddenly being faced with record structures permitting up to 99 victims and offenders each—and intricately annotating victim-offender relationships between each possible pair. (We discuss a further complication associated with inflexible structures—NIBRS's historic tendency to rely perhaps too heavily on automated "edit checks" of input data—in Section D.2.[2]) As tough as NIBRS data records might be to populate, they were also not overly amenable to analysis after the fact—again, particularly in the earliest days of NIBRS development. Lost opportunities for building the case for wider NIBRS adoption were a consequence. The design and structure of NIBRS are among the reasons we argue for viewing full-participation NIBRS as an intermediate step to a more fully modernized attribute-based crime reporting system; it is, ultimately, a database structure developed in the 1980s and not scrutinized for operational effectiveness. This barrier is also partially mitigated—at least with regard to post hoc analysis—simply by advances in computing speed and power since the 1980s. At the same time, NIBRS coordinators may also be able to leverage and build from the work of the Inter-university Consortium for Political and Social Research (ICPSR) at the University of Michigan, which has carried out work to manipulate NIBRS structures into "flat files," or more conventional record layouts, to make them more amenable to analysis. A final mitigation structure is to continue the work that has already been done within the FBI's UCR Modernization program in making UCR and NIBRS more usable for agencies and stakeholders alike, including the development of additional data-input format options (e.g., in XML) and the evolving Crime Data Explorer portal for dissemination.

A final major barrier is related to the cost-benefit assessment that underlies the generation of call or incident reports by law enforcement officers in the first place; it is the *perception that more detailed, NIBRS-type reporting would take more time than "what we do now" in local departments, keeping officers from their duties on the streets and writing reports instead.* The adage typically attributed to Josiah Stamp in 1929, that "the Government are very keen on amassing statistics," but "what you must never forget is that every one of those figures comes in the first instance from the [village watchman],[3] who just puts down what he damn pleases," is grounded in some truth. But so too is a theme that recurred throughout the panel's discussions: It would also be a major error to conceptualize officers' reports purely in the abstract as "data input" cogs in the system and to view officers as report writers who happen to have the power of arrest. Adding to the tension, as we have previously discussed, is that the same officer report that is the germ of police-report data can also be a sensitive legal document—the precursor to the investigative process and the documentation of

[2]A related, sometimes expressed barrier related to relatively inflexibility in NIBRS standards is *concern that the process of becoming "certified" by the NIBRS program to submit data in the NIBRS format is overly time-consuming.* Indeed, it has been stated that failure to achieve all of the requirements for NIBRS certification is the major single factor that has held up direct connection between the Defense Incident-Based Reporting System (DIBRS) and NIBRS (U.S. Department of Defense, Inspector General, 2014).

[3]The quotation—referring to the collection of information in India—used the term *chowty dar.*

attempted and completed offenses that may be held up to sharp scrutiny in the judicial process.

This is a perpetual concern and one without a pat, easy remediation strategy; for local departments, the time management trade-off between an officer spending even 20 minutes writing out a report versus returning to active patrol after a quick half-page scribble is neither easy nor obvious. We took one important mitigation step in formulating our proposed classification and associated attributes; as described in Section 5.2.2 of Report 1, we very deliberately limited the range of attributes to things that would be relatively straightforward for a beat officer (or survey respondent, or administrative database recordkeeper) to objectively assess and document. In doing so, we excluded a great many things that would be useful or even important to know about crime offenses and kept the scope of the attribute collection to roughly the same level as the current NIBRS data elements. But, otherwise, the only mitigation of this concern can come through experience. Departments will rethink and retool their own procedures for reporting (and records management based on those reports), work through any complications that may arise, and build from lessons learned in pilot work and by NCS-X/NIBRS adopters adjusting to new work flows. In our discussions, the ideal for initial report generation that recurred throughout was, effectively, a software-solution model—one that, like tax preparation software, echoes concepts of good survey design by guiding people through simple, stepwise questions. In this structure, an officer might characterize a broad, general offense group, but the proper offense code(s) would ultimately be derived from the attributes and other information recorded in the report.

This remains a useful ideal vision that is likely already achieved in part in some agencies' information management systems. But just as information systems vary across departments, so too will agencies' proximity to this kind of ideal division. Officer report generation (and the priority that local departments put on it) runs the gamut from hand-filled paper "top sheet" descriptions (with or without a dictated narrative) that may be keyed later by records clerks to records fully born in electronic form in officers' Mobile Computer Terminal units in their vehicles. We are heartened through our discussions with practitioners, and those who have gone through the conversion to fully electronic reporting, that this is a general area in which local agency culture is continuing to evolve: Our sense is that report writing is increasingly seen by local law enforcement agencies as an indicator of their quality and effectiveness, and so improving it and training for it is growing in importance (ideally, with generation of the material for crime statistics as a transparent by-product). When we asked our meeting participants about the difficulty in making the transition—whether there were loud complaints about the time investment and such—the collective answer was that there was early resistance to change (just as there is with any policy), and that the transition was easier for newer personnel who were less steeped in older routines. Our discussions suggested other management strategies that might be helpful—striving always to make reporting as painless as possible (e.g.., tailoring the reporting "form" to the data entry mode or minimizing the number of "clicks" needed to complete fields if a report is being completed on a mobile device), emphasizing the role of supervisors in encouraging accurate reporting and catching errors, and rewarding successes (e.g., when a crime is solved or case is won based on precision in reporting, making those successes known). But, in the end, the only way to overcome resistance is by simply insisting on the policy change and just doing it.

D.2 ENSURING DATA QUALITY

As recounted in detail in Section B.4, data quality is a topic with which the UCR police-report data apparatus has struggled for some years. Researchers have long documented shortcomings in the UCR data, including how the UCR Program deals with missing data through processes like imputation. The first serious attention to the FBI's imputation routines appears to have come about when BJS was called upon to develop formulas for allocating Local Law Enforcement Block Grant Program funds—tied to UCR totals—by law in 1994. Following the law's guidelines, BJS analyzed raw crime data provided to the national UCR program, but quickly found that 19 percent of the 18,413 "contributing" agencies in the relevant time period (36 months, 1992–1994) had not reported crime counts for *any* month in the range; an additional 17 percent missed at least one month of reporting (Maltz, 1999:7,9). Thereafter, BJS provided some technical assistance to the FBI on improving its imputation processes, including convening an expert conference on the topic.[4] Concern over imputation routines—for UCR Summary counts, much less NIBRS-format data—exists for all offenses.[5] To the extent that the FBI imputes for nonresponse now, it seems to be heavily if not exclusively based on stratification along a single variable—population of the agency's service population—which is overly simplistic; partial-year counts are commonly weighted up to a full year without reflecting any kind of seasonal component.

There is an unfortunate tendency to assume that simply getting information into a records management system is sufficient or that the process of entering

[4]In 2004, BJS issued a technical report (Bauer, 2004) summarizing BJS's final allocation formula and the amounts of money dispersed to the statement under the Local Law Enforcement Block Grant Program from 1996 to 2004. BJS was formally tasked with deriving the formula for the Edward Byrne Memorial Justice Assistance Grant program, the successor to the block grant program (Hickman, 2005).

[5]The problem may be particularly acute for the offense of arson; though arson was added to the UCR in 1979, the extent of missing data in the arson series has been such that subnational estimates for arson have not been produced.

data into a system with numerous "edit checks" and the running of automated validation routines is adequate to alleviate error. It is still very much possible to have data submissions that meet all technical requirements and that pass any number of procedural edit checks but still suffer from major quality problems. At the national UCR-program level, the existing routines that have been deployed to detect problems in UCR Summary data (Akiyama and Propheter, 2005) are reasonable approaches:

- Cross-sectional outlier detection, or comparing an agency's report to similar agencies based on population size, urbanization, agency type, and geographic location;
- Longitudinal outlier detection, or detecting seriously aberrant counts in an agency's reports over time; and
- Proportionality outlier detection, or examination of deviations from stratum distribution on a number of issues (weapon distributions; proportion of simple and aggravated assault; distribution of monetary loss in property crimes; proportion of violent to property crimes).

However, these checks are limited to outlier detection—finding numerical anomalies but not necessarily data quality problems due to misapplied definitions or the like. From our conversations with law enforcement practitioners, we heard multiple examples of agency staff who, pressed for time and under demand to show results in getting records entered into the system, responded to automated edit-check flags by simply changing relevant codes to "NULL"—enough to satisfy the edit check, but decidedly inaccurate. A U.S. Department of Defense, Inspector General (2014) report, exploring reasons why the Defense Incident-Based Reporting System (DIBRS) was not yet certified to contribute data to NIBRS, made the jarring discovery that at least one police agency within the Defense Department had effectively formalized the practice of entering "null" codes in DIBRS records that had been flagged as erroneous/anomalous by DIBRS edit checks. Functionally, this process had the effect of correcting errors through deletion of the data. Meanwhile, NIBRS itself—with its own numerous edit checks—has adopted a strategy about which we heard much from numerous stakeholders at our meetings: a curiously unforgiving "all or nothing" approach under which entire NIBRS incident records may be discarded if a relatively minor edit check is failed. The quintessential example is a homicide that occurs in conjunction with a robbery, in which the pure edit-check approach would reject the entire record if the "value of stolen property" element associated with the robbery is missing/out of bounds. This default approach, if not revised and resubmitted, would delete the record of a homicide because the value of a watch stolen in the course of the robbery was not properly recorded.

In the past, and continuing today, there is no real substitute—for purposes of finding the root causes of data quality problems—for using multiple quality evaluation methods, because no one methodology can assure quality data. The most common way for this to occur is through direct audits, and the matter of audits is one in which our conversations with law enforcement executives suggested considerable esteem. However unusual it is to hear active requests for an audit of any kind, our discussions continually included the expressed hope that audits from the national UCR and state programs would happen more frequently and more intensively. Local departments that have experienced public challenges to their credibility over alleged miscounting of crime are particularly appreciative of the bolstered trust and accountability that can result from the audit process. Historically, the FBI's process of audits—the Quality Assurance Review (QAR) process—has been conducted with the states on a triennial basis. Roughly speaking, the UCR Program would develop a list of agencies to be covered in the audit; this would be negotiated with the state program and, typically, 10–12 agencies would be selected for inclusion in the audit. Records and reports drawn from agency files would then be reviewed by FBI staff, either in person or remotely via mail. Budget constraints promoted a hiatus in the FBI's QAR process, but the audits were set to begin again in mid-2016 in somewhat expanded format, based on feedback from a focus group of state UCR managers. Going forward, the FBI CJIS Audit Unit was set to offer three types of audit/review (Criminal Justice Information Services Division, 2016a:9–10):

- The successor to the traditional FBI spot-audit, a Statistical/Quantitative Review, would continue on the triennial cycle; "the national UCR Program staff will identify two offenses to be reviewed as of October 1 of each triennial cycle"; "up to 400 incidents per each offense" would be reviewed by FBI staff on a "simple random sample basis" to identify audit targets. For those sample incidents, the relevant records (such as case files and narrative reports) would be compiled by the state UCR program and reviewed by FBI staff "over a three-day period" at the state agency.[6]
- A Service/Qualitative Review would be an "optional add-on" to the first audit type, in which the state UCR program could designate up to two local agencies within the state to be audited—in the same manner as the Statistical/Quantitative Review—by FBI staff.
- Special Reviews would be permitted on a case-by-case and available-resource basis, and are audits done on an ad hoc basis off of the triennial cycle for "situations that require immediate FBI assistance."

[6] A later quarterly program update noted that a Probability Sampling Method had been added to this review, by which "the CAU staff will evaluate the local agency records from a sample of local agencies" in cases where "a state program is unable to collect local agency records in support of the Statistical Review methodology," but the process was not clearly explained (Criminal Justice Information Services Division, 2016b:7–8).

There are any number of impediments to full, complete, and accurate reporting of crime and offense incidents by local law enforcement agencies—some malignant, some benign, and all potentially corrosive to an agency's credibility. Some of these include the following:

- Varying standards across agencies for dealing with incidents, including what triggers completion of a report (i.e., whether citizen complaints/calls for service are routinely logged and start the reporting process versus having to physically report offenses at station);
- Design and implementation factors in the information management systems used at local level, including ease (or difficulty) of data entry for acceptance in system;
- State of police–community relations in the locality (poor relations may result in fewer crimes being reported to authorities);
- Operational siloes and rivalries within the agency that may impede data sharing, as may rules on victim confidentiality (e.g., domestic violence or sexual crime units may not routinely share data with records clerks);
- Inadequate training in report preparation or system usage; and
- Inaccuracy through inability to meet state or national data submission deadlines, including through shifting staff/resource priorities or software/network failures.

In addition to audits by either the national or state level, the other way in which multiple sets of eyes might be drawn to potential data quality problems is through the institution by local departments of multistage processes for the review of officer-filed reports—that is, trying to catch sources of error at the origin of the eventual crime data record. Multistage review, in which both a supervisor and records personnel would have the ability to suggest corrections—and the report/record would be opened for correction and resubmission—was expressed to the panel as an effective best practice model to consider. But it is also one, it was just as quickly noted, that smaller and less-resourced departments might simply be unable to do or afford (not least of which because dedicated, separate records staff might be beyond the agency's resources).

Against the backdrop of all these issues, it is very deliberate that the ideal coordination and governance roles that we outlined for national crime statistics repeatedly reference data quality issues. There is much work to be done in bringing procedures for handling missing data up to expected standard practices of a statistical agency—and likewise for the development of edit checks and remediations that flag anomalies but that do not risk destruction of important partial data. Part of demonstrating national commitment to the collection of national crime statistics must be finding ways to continue to improve and expand systematic spot audits. And, in reframing national crime statistics data collection as a strong federal-state cooperative program, it should be possible to converge on something close to this ideal, rough workflow for the police-report component:

- The national coordinating body/governance structure establishes uniform, national definitions of offenses (via the classification), supporting rules, and data quality procedures.
- Local law enforcement agencies submit offense reports (or transfer records) on a monthly basis (at least) to a state-level coordinator, which is enabled to serve as a check on data quality: vetting the data, converting codes (if not already done by local software) in records to meet the national definitions, flagging incident reports/records with errors, and returning them to the local agencies.
- Local agencies correct errors and resubmit to the state coordinator; data are checked again and, pending successful review, forwarded to the national coordinating body.
- The national coordinator vets the data, flagging incident reports/records with errors, and returns them to the state coordinator; in turn, the local agency is contacted and corrects and resubmits problematic reports.
- On a periodic basis, local agencies take stock of recurring problems, retraining data entry personnel as necessary and modifying software (or contacting vendor to modify software) to resolve issues.
- On a periodic basis, state coordinators take stock of recurring problems, modifying software (or contacting vendor to modify software) as needed to resolve issues.

In this climate, we think local agencies would realize direct benefits—to their credibility and accountability in the eyes of the populations they service—from a quality-oriented approach, and that this would make it easier to sustain the case for active participation in the national program.

In closing this section, we also wish to at least tentatively mention that new technologies may open additional avenues for improving overall data quality. Third-party triangulation is one approach that is done on a limited basis now, effectively matching offense-specific police-report data to incidents identified by other sources such as advocacy groups (e.g., for hate crime, domestic violence, and sexual assault cases). Such an approach might be applied for offense types for which incident-level data are available in both police-report data and from administrative/external resources. But, more generally, Web-scraping approaches that touch a multitude of Internet pages open up the possibility of such triangulation for some traditional and new offense types. In 2016, BJS briefly advanced a revision to its Arrest-Related Death reporting program that basically used Web-scraping (scouring open-source/media sites for reports of use-of-force incidents) that could then be compared to law enforcement agency-supplied data—essentially, asking local agencies how and whether they accounted for a specific sample of incidents in the media-source data. As we noted in Section 1.3 (Chapter 1), we do not believe it to be our role to comment on use-of-force data and their development in this report, but the methodological idea is a powerful one; done well, it amounts to the dual-systems estimation technique that is now a principal source of estimation of differential undercount or overcount in the decennial census. It could be a very interesting methodology to study quality of data on sensitive offenses such as bias-motivated or hate crimes.

D.3 PUTTING CRIME IN CONTEXT AND ENABLING EFFECTIVE COMPARISONS

One of the historical themes in the development of national crime statistics that we noted in Appendix B was the escalation of warnings against simple, direct comparison between multiple city-level counts or rates of crime published in the UCR tables. The warnings were, and are, grounded in truth. A variety of external or contextual factors may partially explain differences in the figures across jurisdictions and over time, and so a simple difference in rates is not necessarily indicative of differences in criminality. In our panel's workshop sessions, we heard from local agencies that appreciated the intent of the admonition against comparison and, indeed, had been burned in the past by overly simplistic comparisons drawn by city officials, by reporters, or by members of the public. Budgetary and resource assumptions were based on comparisons between jurisdictions similar to each other only (at best) by total population, ignoring any sociodemographic or geographic context. Yet, at the same time, it has long been recognized that benchmarking—comparison of how a particular agency's offense patterns compare with peers—is the main (if not sole) purpose and perceived benefit of nationally compiled crime statistics to many crime data users/stakeholders. In our panel's interactions with crime data users, local agencies noted the inherently contradictory nature of typical UCR data presentations: extensive tabular listings that practically beg the drawing of direct comparisons while simultaneously arguing against doing so.

An underlying problem is that inappropriate comparisons are often drawn, out of proper context, precisely because it is not easy to get the kind of supplemental, auxiliary data that *could* put crime figures in context and permit valid comparisons. In our discussions, we heard of local departments calling and working with their peers to identify jurisdictions that are similar with respect to poverty rates, extent of segregation, or other sociodemographic factors. Agencies and other data users that are more savvy to this process of digging for context wind up relying on U.S. Census Bureau data almost as heavily as the crime statistics themselves. Arguably, as we heard from other data users, it does not help matters that the exact manner by which the FBI computes the

Box D.1 Framework for Measuring Crime and Public Safety proposed by the Crime Indicators Working Group

In 2013–2015, the Bureau of Justice Statistics (BJS) convened an ad hoc Crime Indicators Working Group (CIWG) comprised of representatives of various law enforcement agencies and stakeholder groups. Accepting the premise that UCR Summary–type measures were no longer adequate to meet user or community needs, the CIWG discussions called for combining measures based on incident-based crime reporting with a variety of contextual variables to put crime numbers in community context. Specifically, the group suggested that measures be generated in five high-level categories, as follows:

1. **Victimization**, to be further subdivided by crimes against persons; crimes against commercial establishments and noncommercial organizations, and indicators of public safety and disorder:
   - The crime-based victimization measures are intended to include both offense counts and population-based rates, with those counts and rates disaggregated by: victim demographics, victim-offender relationship, weapon use, location (including locationless/cybercrime), and whether incidents involved multiple victims or offenders.
   - Some of the indicators of public safety and disorder might be derivable from the National Crime Victimization Survey or other extant data; others might require a separate survey.
2. **Offenders**: Counts and population-based rates of arrest, disaggregated by such arrestee characteristics as demographics, alcohol and/or drug use, computer involvement in facilitating the crime, and arrestee weapon use (including discharge of a firearm during incident).
3. **Temporal and spatial patterning of crime**: Premised on the collection of geocoded crime incidents; also includes collection and analysis of general crime location type (private or public location), time of day, and urbanicity (urban, suburban, rural).
4. **Harm**: Counts of crimes resulting in injury, by type of injury; estimate of money or property lost by both persons and commercial establishments through crime; measures of extent to which crime victims seek assistance, including percentage of crime victims reporting the incident to law enforcement.
5. **Response to crime**: Measures of the function of other stages in the criminal justice process, including count of crimes cleared; arrests resulting in prosecution; number of filings and outcomes/sentences by offense in court; admissions to correctional supervision/custody and number of persons returning to community.

SOURCE: Presentation by Howard Snyder, Bureau of Justice Statistics, to the panel on May 11, 2015, and participation by panel members and staff in CIWG sessions

agency service population totals that serve as denominators for calculated rates is fairly opaque to the public.

Concurrent with the beginning of our panel's work, BJS convened several meetings of a Crime Indicators Working Group (CIWG) comprised principally of local law enforcement executives and their support organizations, to which the basic question was put: What information about crime or related to crime could BJS supply that would be most useful to local law enforcement, not only in day-to-day operations but in engagement with members of their community? The basic, ideal indicator data framework developed by the CIWG in its discussions is shown in Box D.1. Not surprisingly, CIWG participants expressed wishes for fine-grained crime data that could support studies of the temporal and spatial patterning of crime—ideally at the level of city neighborhoods, if not finer. Relative to response to crime, the group also saw a need for more longitudinal analysis of outcomes through later stages of the justice system (i.e., from arrests through prosecution/court filings through correctional disposition) in order for local agencies to measure their ultimate success or effectiveness in addressing crime problems. But the brief framework description in the box understates—in its reference to "indicators of public safety and disorder" as well as community sociodemographic information—the strong desire expressed by the group for the simultaneous display and dissemination of crime data and the contextual/noncrime data that could help explain variations therein.

Accordingly, this section appeals for a change in approach in the presentation of modern, national crime statistics: Give data users the necessary tools to make crime rate comparisons in proper context rather than dismissively advise against doing that one thing—drawing comparisons—that is the primary use of nationally compiled data for most stakeholders. Specifically, we suggest that the dissemination platform for new, national crime statistics should also provide straightforward access to auxiliary data resources and tabulations that may permit meaningful comparisons between jurisdictions and across time. The coordinating body for crime statistics that we recommend in Chapter 3 should work with the U.S. Census Bureau, possibly through the quinquennial Census of Governments and related geographic support services, to articulate (and update) geographic boundary files for law enforcement agency service areas—making it easier to generate special tabulations (of population estimates and data from other sources such as the American Community Survey) for more accurate rate calculation and fuller analysis of crime data in community context. To clarify and reiterate, this is not an argument that the national crime statistics program should be in the business of creating place-based reports and analysis on its own, but that it should make the construction of the same—by other stakeholders—feasible. As hinted at by the CIWG in its proposed framework, the generation of contextual variables on the community context of crime—relevant noncrime data that may usefully bear on crime levels and trends—will likely involve multiple sources. Accordingly, an important frontier for the NCVS is making more effective use of the survey's completed interviews—including the survey's "screener" portion and those questions asked of all households whether they have specific victimization incidents to report or not—to generate this kind of information.

D.4 BUSINESSES AND ORGANIZATIONS AS ACTORS IN CRIME STATISTICS

A consequence of decades of heavy focus on a small number of violent and property crimes—and near-exclusive focus on interpersonal crime—is that the nation's current crime statistics are particularly poor at addressing crime in which businesses, organizations, or government agencies are actors in crime as victims or offenders. A major underlying cause of difficulty in understanding commercial victimization illustrates additional challenges. For a business to admit that it has been the victim of a major credit card hack or extensive organized theft is an exposure of vulnerability and weakness that may put them at competitive disadvantage with their peers. Big businesses and large retailers tend to have their own security and investigative personnel and their own protocols for sharing information on offenses; they may report some offenses (e.g., known shoplifting incidents involving goods above some threshold value), but otherwise, crime reporting may expose too much vulnerability. Hence, where others see "crime" and "policing," businesses see "risk" and "risk mitigation"; where others see "theft," "pilferage" (theft by an employee), or "shoplifting," businesses see "stock loss" or "shrinkage." The picture of crime in the United States is necessarily incomplete without more attention to businesses or non-person actors; it might be convenient to just write off shrinkage, or fraud against the government, or like offenses as seemingly invisible offenses, but the sums of money involved are immense.[7] Ultimately the costs associated with these offenses show up in consumer prices or charges to the taxpayer downstream.

Because the issue has parallels with crime collection, a brief summary of efforts to assess retail theft is noted:

- Federal law enacted in January 2006 (119 Stat. 3092) mandated that "the Attorney General and the Federal Bureau of Investigation, in consultation with the retail community," create a national database "to track and identify where organized retail theft[8] type crimes are being committed" in the nation. The law specifically directed that the database be "housed and maintained in the private sector," but that it "allow Federal, State, and local law enforcement officials as well as authorized retail companies (and authorized associated retail databases)" to populate

[7]For example, Sparrow (2000) suggested that health care fraud alone, as a part of fraud against the government, amounts to hundreds of billions of dollars annually.

[8]The law defines "organized retail theft" as (a) violations of state theft/shoplifting law that involve both "quantities of items that would not normally be purchased for personal use or consumption" and "the purpose of reselling the items [or] reentering the items into commerce"; (b) the receipt, sale, disposal, or any handling of property "[known] or should be known to have been taken" in violation of condition (a); or (c) coordination/organization to do either of the preceding things (119 Stat. 3092).

and access the database.[9] The resulting database, the Law Enforcement Retail Partnership Network (LERPNet), was developed through a partnership of the FBI, the Retail Industry Leaders Association, the National Retail Federation, and the Food Marketing Institute (FMI); responsibility for LERPNet was passed to Verisk Retail (part of Verisk Analytics) in 2011, and Verisk announced a third release of the underlying database platform (dubbed LERPnet Falcon) in June 2013.[10]

However, what began as a clear legal mandate for data collection—one with a promise of funds attached, and one explicitly defined to be on business's terms and in private-sector operation—seems to have unraveled. When the U.S. Government Accountability Office (2011) was tasked to report on the state of information on organized retail theft, the discussions with corporate security personnel summarized in the report were fairly scathing of LERPnet. The report found that businesses had found the system difficult to use and were finding considerably greater success in banding together in regional consortia, maintaining their own databases of organized retail theft occurrences. Meanwhile, as of August 2017 (and for some unknown length of time prior to then), the LERPnet database has at least been closed to new participants and is possibly shut down entirely; the website http://www.veriskretail.com/lerpnet/ returns an advisory that "Verisk Retail is not currently onboarding LERPnet clients as we re-envision our [organized retail crime]-oriented solutions for retailers and law enforcement," and other commonly listed URLs for the LERPnet interface appear to be nonfunctional.

- To provide some general information on retail theft, the National Retail Federation has sponsored an annual National Retail Security Survey for just over 25 years; the survey is now administered as a web survey to a small set of company executives or security officers (80–100 in the 2015–2017 administrations of the survey, judging from press releases accompanying the survey results). The 2016 instance of the survey (covering retail results in calendar 2015) estimated inventory shrinkage levels as averaging 1.38 percent of total sales among respondents—extrapolating to a $45.2 billion national problem. The survey tends to show that the majority of that shrinkage is accounted for by two classes of theft—external theft (including shoplifting and organized retail theft,

[9]The law also called for creation of a joint task force to "combat organized retail theft," and so the $20 million in authorized appropriations over four fiscal years was aimed principally at "educating and training federal law enforcement" and "apprehending and prosecuting individuals," with the creation of the database as the third and final priority. The law included more generic assertions that the Justice Department "make available funds to provide for the ongoing administrative and technological costs to federal law enforcement agencies participating in the database project" and that it "may make grants to help provide for the administrative and technological costs to [participating] State and local law enforcement agencies."

[10]See http://www.verisk.com/archived/2013/verisk-retail-launches-lerpnet-falcon.html.

at 36.5 percent in the 2017 survey) and internal theft (pilferage/employee theft, at 30 percent), with administrative or paperwork error, vendor fraud or error, and unknown sources comprising the rest. But, beyond a rough glimpse of the scope of the problem, the survey is not capable of providing additional detail or insights. The survey also asks about the presence of various security measures, but the univariate data from the survey are an incomplete measure of their potential effectiveness.[11]

- Finally, and directly akin to this panel's work in trying to define the scope of crime in general: The Retail Industry Leaders Association (RILA) commissioned a study to go back to first principles in addressing the problem of retail loss generally. The resulting study, *Beyond Shrinkage* (Beck, 2016), does exactly what this study did as its first step: It defines a hierarchical classification structure that decomposes the concept of "total retail loss" as a prelude to suggesting pilot work and data collection to begin to fill in that classification with data. Specifically, the proposed taxonomy imposes a first-level split by retail setting (store, retail supply chain, e-commerce, and corporate); within each of these retail settings, known stock loss is further decomposed into final categories depending on whether they are malicious or nonmalicious in nature (each of the four settings are subject to unknown stock loss, as well). The full details and definitions of the classification are given in the study report but, illustrative of the range of crime types encompassed in this framework, the general malicious categories under the four retail settings are:
  - In-store: external theft, internal theft, customer frauds, voucher/loyalty card scams, burglary/criminal damage/arson, cash theft
  - Retail supply chain: internal theft, burglary/criminal damage/arson, in-transit losses
  - E-commerce: customer frauds (including charges to false/stolen credit card or fraudulent claims of failed delivery)
  - Corporate: corporate frauds

We draw two principal messages from our discussions with stakeholders and deliberations and from the preceding information on measuring retail loss. The first is that, fundamentally, collecting data on crimes affecting businesses largely amounts to trying to achieve information sharing in a culture where information sharing is anathema. Hence, to our system's data collection

[11]The full survey results are available to National Retail Federation members, but the basic shape of the surveys can be inferred from federation and related press releases, such as https://nrf.com/resources/retail-library/national-retail-security-survey-2017, http://losspreventionmedia.com/insider/loss-prevention/2017-national-retail-security-survey-now-available-for-data-submission/, and https://nrf.com/media/press-releases/retailers-estimate-shoplifting-incidents-of-fraud-cost-44-billion-2014.

modes of police-report data, personal/household survey information, and administrative-type records, it is likely that a fourth option involving the cultivation of "safe havens" for information sharing between organizations may need to be developed. The idea is that the shared risk of revealing and discussing crime in such a safe haven would be offset by the shared reward of picking up best practices for responding to offenses. Such safe havens might engender opportunities for anonymized data collection more effectively than police-report data or commercial victimization survey data ever could, though it may take time for culture change—and trust—to develop. One possible model here is the National Cyber-Forensics and Training Alliance (NCFTA), a Pittsburgh-based nonprofit with whom we spoke about the challenges of defining cybercrime but that also provides a useful, broader organizational example.[12] NCFTA effectively serves as a roundtable for companies, government agencies, and academia to work together to combat cybercrime. Its work involves identifying individual and collective risk, formulating strategies for mitigating cybercrime and stopping its thread, and ultimately neutralizing it (through law enforcement, as in efforts to shut down several "dark web" sites, or through shared software tools and solutions). Primarily, the alliance is meant to serve as a bridge for both resources and information—getting partners to band together and share information to make difficult problems tractable, for the good of the whole coalition, when individual companies' risks/exposures might not be worth the cost of seeking law enforcement intervention. The second principal message is the related point that trade associations such as RILA *might* emerge as the source of data and key indicators for particular offenses, on the grounds that businesses *might* be more amenable to information sharing within an association than with law enforcement, survey researchers, and the like. Moreover, enterprises by trade support organizations such as the proposed classification system for total retail loss should be strongly encouraged, fostering principled stances to define and measure things related to crime.

[12]In addition to general cybercrime, NCFTA currently does major work in the areas of cyberfinancial crime, consumer/e-commerce fraud, and malware deployment.

– E –

Excerpted State Legal Requirements for Crime Reporting

In this appendix, we attempt to summarize the relevant legal context for crime reporting in each state. To the extent practicable, the listing for each state summarizes or quotes from state statute or other documents, covering 5–6 items:

- The opening bullet point tries to identify the most basic mandate for crime reporting—typically, the law from which the state's Uniform Crime Reporting (UCR) Program draws its base authority. (The membership listing of the Association of State Uniform Crime Reporting Programs, at http://asucrp.net, was consulted to start the listing of programs.) The maintenance of criminal history record databases or development of intrastate communication networks for law enforcement use are among the things that are typically dealt with in parallel passages of state codes; we have tried to avoid citing these (and instead focusing more on pure "crime statistics") except in cases where the state UCR program appears to draw authority from such broader initiatives. The first bullet point in each state's entry below also attempts to indicate *whether local law enforcement agencies are legally required to transfer crime-report data with a state-level entity* or whether participation is voluntary (and merely encouraged). A common structure in several states is the base appearance of voluntary contribution—stressing "cooperation" between the local agencies and the state—followed by a statement of penalties for nonreporting; we tend to classify these constructs as mandatory response.

- A second bullet point indicates whether *transfer of data on criminal offenses (typically from the state level entity) is required by state law*, whether it is simply encouraged (i.e., the law only obliges the state agency to "cooperate" with the national UCR program), or whether the law is apparently silent on the issue.
- A third bullet covers the *status of incident-based reporting in the state*—notably, whether the state maintains its own incident-based reporting system or whether it is certified by the Federal Bureau of Investigation (FBI) to contribute National Incident-Based Reporting System (NIBRS)-format data. This bullet typically draws from four sources: (1) the "JRSA inventory" of incident-based readiness, a set of profiles assembled by the Justice Research and Statistics Association (JRSA) at http://www.jrsa.org/ibrrc/background-status/nibrs-states.html (though no date is given on the profiles and some of the information might be dated); (2) the BJS National Crime Statistics Exchange (NCS-X) project page at https://www.bjs.gov/content/ncsx.cfm, including its link to the press release detailing recipients of NCS-X grant monies (https://www.bjs.gov/content/pub/press/NCS-X16pr.cfm); (3) the state UCR program's website or the most recent annual "crime report" published by the state for information, as appropriate; and (4) the table "Offense Type by Agency, 2015" from the FBI's 2015 report of NIBRS data (published at https://ucr.fbi.gov/nibrs/2015), which summarizes data from "all law enforcement agencies that submitted 12 months of data to the UCR Program via NIBRS and whose data were also converted to summary data for publication in *Crime in the United States*, 2015" (per the associated Data Declaration for the table).
- Fourth, we indicate whether *a formal, annual report of crime statistics is required* by law or whether the law is apparently silent on the issue.
- An optional fifth bullet includes discussion of *other key features evident from the state's crime reporting laws*—an admittedly incomplete inventory, as our search was generally constrained to "neighboring" sections of state law rather than a full scrutiny of the code where related topics might appear.
- The final bullet identifies the *Statistical Analysis Center (SAC)* designated by the state; the SACs receive some funding through Bureau of Justice Statistics grants and commonly serve as analysts and "interpreters" of crime and justice data for state policy makers. The SACs vary greatly in ther basic legal statute. Some are colocated with the state UCR program, others are in a different government agency, and some are either attached to academic institutions or are operated by nonprofit entities. For this entry, we used the JRSA membership list as a base.

Box E.1 State Uniform Crime Reporting Programs by Incident-Based Reporting and National Crime Statistics Exchange (NCS-X) Funding Status

Fully Incident-Based/NIBRS Compliant (NCS-X ineligible):

Arkansas	Colorado	Delaware	Iowa
Idaho	Kentucky	Michigan	Montana
North Dakota	New Hampshire	South Carolina	South Dakota
Tennessee	Virginia	Vermont	West Virginia

At Least 80 Percent NIBRS Compliant (NCS-X ineligible):

Connecticut	Massachusetts	Ohio	Rhode Island

State-Led (deemed to not require NCS-X funding for expansion/implementation):

Georgia	Indiana	Oregon	Texas

Awarded NCS-X Planning Grant in Fiscal Year 2015 or 2016:

California	Florida	Illinois	Maryland
Nevada	Wisconsin	Wyoming	

Awarded NCS-X Implementation Grant in Fiscal Year 2015 or 2016:

Alabama	Hawaii	Kansas	Louisiana
Maine	Minnesota	Missouri	North Carolina
Nebraska	New Jersey	New York	Oklahoma
Pennsylvania	Utah	Washington	

Not NCS-X Awarded (as yet):

Alaska	Arizona	New Mexico

No State UCR Program: Mississippi

SOURCE: Adapted from Bureau of Justice Statistics NCS-X page, https://www.bjs.gov/content/ncsx.cfm

As a convenient summary of the status of incident-based reporting in the states, Box E.1 derives from the BJS NCS-X project page and lists the state UCR programs by their eligibility status within the NCS-X program (e.g., not eligible for NCS-X funds because already fully NIBRS compliant) and type of NCS-X grant awarded (e.g., Planning, for from-scratch development of requirements, versus Implementation, for refinement or expansion of an existing incident-based recording system).

ALABAMA

- *Local law enforcement agencies required to submit crime statistics to state entity:* The Alabama Criminal Justice Information Center (ACJIC), within the Alabama Law Enforcement Agency, is required to "provide for a uniform crime reporting system for the periodic collection and analysis of crimes reported to any and all criminal justice agencies within the state" (Code of Ala. § 41-9-620). § 41-9-631 of the code further provides that the "Alabama uniform crime report" is to "contain the number and nature of offenses committed, the disposition of such offenses and such other information as the commission shall specify relating to the method, frequency, cause and prevention of crime." § 41-9-631 makes reporting of this information mandatory for "all criminal justice agencies within the state;" submission of data by "any governmental agency" with relevant information but that is not considered a "criminal justice agency" is voluntary (§ 41-9-632), but ACJIC is obliged to include such voluntarily provided data in its statistical reports.

- *Transfer to FBI/national compilation encouraged by law:* § 41-9-621 directs the commission to "cooperate with other agencies of this state, the crime information agencies of other states and the uniform crime reports and national crime information center systems of the Federal Bureau of Investigation or any entity designated by the federal government as the central clearinghouse for criminal justice information systems in developing and conducting an interstate, national and international system of criminal identification, records and statistics."
- *State maintains own incident-based reporting system:* According to the brief history section in recent *Crime in Alabama* volumes (e.g., Alabama Law Enforcement Agency, 2016), ACJIC began offering a state Incident-Based Reporting System as a strictly voluntary option for agencies in the state. Said system "also provides a self-contained internal record-keeping system for the agency." Though voluntary, ACJIC indicates that "more than 99 percent of Alabama's law enforcement agencies submitted Incident Reports" in 2015, out of the "approximately 350 agencies [that] reported known offenses to Alabama's UCR." Further, ACJIC indicates that the state incident-based reporting system has undergone revision to comply with NIBRS standards:

  > In 1986, ACJIC organized a 32-person UCR Committee to review the IBR report forms. This committee was made up of law enforcement personnel from small, medium and large police departments and sheriffs' departments, computerized agencies, the FBI, and the Bureau of Justice Statistics (BJS). The committee reviewed and analyzed the 10-year-old forms and decided enhancements were needed. The IBR report forms were redesigned with input from Alabama's local agencies. In 2006, these forms were redesigned to conform to the FBI's Nation [sic] Incident Based Reporting System (NIBRS). The new forms became the official state forms in 2010.

  Only one agency in Alabama—the city of Hoover—is NIBRS certified and contributed NIBRS data in 2015.
- *Annual crime report required by law:* § 41-9-621 directs the ACJIC to "periodically publish statistics, no less frequently than annually, that do not identify persons," adding that "such information shall accurately reflect the level and nature of crime in this state."
- *State Statistical Analysis Center:* Also housed in the Alabama Law Enforcement Agency.

ALASKA

- *Local law enforcement agencies required to submit crime statistics to state entity:* Alaska Stat. § 12.62.110 (2014) directs the Alaska Department of Public Safety to, among other duties, "provide a uniform crime reporting system for the periodic collection, analysis, and reporting of crimes, and compile and publish statistics and other information on the nature and extent of crime in the state." Such "data regarding crimes committed within that agency's jurisdiction" "shall [be submitted] at the time, in the manner, and in the form specified by the department" (§ 12.62.130). The penalty for nonreporting is that the department is directed to withhold grant funding to local agencies that do not report.
- *State has no incident-based reporting system*, and is listed in JRSA's inventory as one of the states with no plan for conversion to NIBRS. However, a 2010 amendment to § 12.62.130 added the clause that, "at a minimum," local agencies must report "each felony sex offense committed in an agency's jurisdiction"—thus, in principle, mandating incident-level reporting of defined sex offenses as an adjunct to UCR reporting. In light of these circumstances, it is curious to note—with respect to the state's largest city—two contrary signals:
  - The city of Anchorage has periodically conducted analyses of domestic violence incidents, mining incident-based data from the Anchorage Police Department's records management system. The report of the city's Fourteen Year Study of domestic violence (Municipality of Anchorage, 2006) is an extensive analysis illustrating the utility of incident-based data for understanding complex problems; simultaneously, the report bemoans the lack of comprehensive national data for comparison (citing the JRSA inventory of incident-based reporting status in stating that "only six states currently have no formalized plan to report incident-based data"—neglecting to add that Alaska is among those six).
  - The first publication of results from the parallel database on felony sex offenses (McClure, 2016:2) commends the database's "ability to capture data elements for each reported sex offense not currently captured through the UCR program," such as "victim and suspect demographic information, victim/suspect relationship information, weapon type, and location of the assault." The report then makes basic use of all of these details to useful effect in the analysis, but it is obliged to begin with a footnote: "The Anchorage Police Department provides only statutory information for their reported incidents; victim, offender, location, and weapon data are not provided."
- *Annual crime report required by law:* § 12.62.130 stipulates that the department issue a formal report each year "to the governor and the attorney general" covering "the number and nature of criminal offenses committed,

the disposition of the offenses, and any other data the commissioner finds appropriate relating to the method, frequency, cause, and prevention of crime."

- *State Statistical Analysis Center:* Housed at University of Alaska Anchorage Justice Center.

ARIZONA

- *Local law enforcement agencies required to submit crime statistics to state entity:* Arizona law (AZ Rev Stat § 41-1750) vests authority in the state Department of Public Safety to serve as "central state repository" for criminal justice information; this includes to "collect information concerning the number and nature of offenses known to have been committed in this state and of the legal steps taken in connection with these offenses, such other information that is useful in the study of crime and in the administration of criminal justice and all other information deemed necessary to operate the statewide uniform crime reporting program and to cooperate with the federal government uniform crime reporting program." Language passed in 1992 switched the state UCR program from voluntary response to mandatory response, with the new text of § 41-1750 continuing that the "chief officers of law enforcement agencies of this state or its political subdivisions shall provide to the department such information as necessary to operate the statewide uniform crime reporting program and to cooperate with the federal government uniform crime reporting program."
- *Transfer to FBI/national compilation encouraged by law:* Though there is no direct statement obliging the department to transfer data to the national UCR program, the twice-occurring text on "cooperate with the federal government uniform crime reporting program" in § 41-1750 (capitalization in original) is certainly consistent with strongly encouraging submission.
- *State is NIBRS certified, but with limited participation:* Arizona does not maintain its own incident-based reporting system, but did acquire computer hardware and software to build a state-level NIBRS repository in 2003. Accordingly, it was certified by the FBI to submit NIBRS-format data even though very few agencies currently do so. The most recent state crime report indicates that only five of Arizona's local agencies currently submit NIBRS-only data, but the report only names four—the police departments in Apache Junction, Gilbert, and Huachuca City, and the sheriff's office of Yuma County (Arizona Department of Public Safety, 2016:24,26). The FBI's 2015 NIBRS tabulations suggests that only Apache Junction, Gilbert, the city of Yuma, and Yuma County contributed 12 months of NIBRS data for the year. Though the Arizona state UCR program has not received NCS-

X funding, the Mesa and Phoenix police departments were awarded NCS-X development funds in 2016.
- *Annual crime report not apparently required by law,* though one has been published on that cycle since the 1970s.
- *State Statistical Analysis Center:* Also housed in Department of Public Safety.

ARKANSAS

- *Local law enforcement agencies required to submit crime statistics to state entity:* The Arkansas Crime Information Center (ACIC) was created as a separate state agency by law and authorized to construct an information system "restricted to serving the informational needs of governmental criminal justice agencies and others specifically authorized by law through a communications network connecting local, county, state, and federal authorities to a centralized state repository of information." However, the enabling statute (AR Code § 12-12-207) also directs the center to "collect data and compile statistics on the nature and extent of crime problems in Arkansas and compile other data related to planning for and operating criminal justice agencies." A final clause of § 12-12-207 invokes terminology authorizing ACIC to "design and administer ... uniform crime reporting systems [to] be used by criminal justice agencies." Pursuant to § 12-12-209, "it shall be the duty of all county sheriffs, chiefs of police, city marshals, correction officials, prosecuting attorneys, court clerks, and other state, county, and local officials and agencies so directed to furnish the Arkansas Crime Information Center all data required" for its work. A local official "who knowingly fails to comply with this subchapter" or related rule "upon conviction is guilty of a violation and shall be punished by a fine not exceeding five hundred dollars."
- *Transfer to FBI/national compilation not apparently addressed by law.*
- *State maintains own incident-based reporting system and is fully NIBRS compliant:* According to the ACIS website (http://www.acic.org/crime-statistics), Arkansas achieved NIBRS-format reporting by all local agencies in the state in 2009. The site also contains some references to the self-styled Arkansas Incident-Based Recording System, though all the information provided on the site relates exclusively to the national NIBRS program.
- *Annual crime report required by law:* § 12-12-207 authorizes the ACIC to "periodically" publish statistical information on crime, but a later section (§ 12-12-217) is date-specific—"on July 31 of each year the [ACIC] shall submit an annual report to the Legislative Council." The law requires two things of the report—that it "[show] the number of persons arrested for each criminal offense classification, comparing the state and each individual reporting agency," and that it "include a breakdown by race of all persons arrested in each criminal offense classification."

- *Other key features:* One of a few states to directly authorize audits of the crime data submitted to the state coordinator, Arkansas goes a step further by directing the director of ACIC to "appoint special information services agents." Said agents, "after proper and sufficient security clearances and training," are directly commissioned to monitor and audit "all records and information" at ACIC (§ 12-12-210). Arkansas is also distinctive for including in the section of state code related to crime information a passage obligating "all physicians, surgeons, hospitals, druggists," or other service providers "that render first aid treatment to a person" to report incidences of gunshot, knife, or burn wounds. Specifically, the law requires reporting of knife or gunshot wounds that "appea[r] to have been intentionally inflicted," burn wounds "that could reasonably be connected to criminal activity," any "second or third degree burn to five percent [or] more of a person's body," or "a burn to a person's upper respiratory tract or laryngeal edema due to the inhalation of super-heated air." These reports are to be filed with local law enforcement or fire department authorities, and the law immunizes reporters of these wounds "from any civil or criminal liability that might otherwise be incurred or imposed" through disclosing the wounds (AR Code § 12-12-602).
- *State Statistical Analysis Center:* Same entity, Arkansas Crime Information Center.

CALIFORNIA

- *Local law enforcement agencies required to submit crime statistics to state entity:* California Penal Code § 13020(b) is arguably the nation's most exhaustive statute in terms of identifying entities that are required "to report statistical data" on crime to the state Attorney General. Specifically, "it shall be the duty" to do so for:

  > every city marshal, chief of police, railroad and steamship police, sheriff, coroner, district attorney, city attorney and city prosecutor having criminal jurisdiction, probation officer, county board of parole commissioners, work furlough administrator, the Department of Justice, Health and Welfare Agency, Department of Corrections, Department of Youth Authority, Youthful Offender Parole Board, Board of Prison Terms, State Department of Health, Department of Benefit Payments, State Fire Marshal, Liquor Control Administrator, constituent agencies of the State Department of Investment, and every other person or agency dealing with crimes or criminals or with delinquency or delinquents, when requested by the Attorney General.

- *Transfer to FBI/national compilation required by law:* "It shall be the duty of the [Department of Justice] to supply, at their request, to federal bureaus or departments engaged in the collection of national criminal statistics"

Box E.2 California Uniform Crime Reporting Requirements as Amended by OpenJustice Data Act of 2016

Section 13010 of the Penal Code is amended to read: **13010.** It shall be the duty of the [California Department of Justice]:

(a) To collect data necessary for the work of the department from all persons and agencies mentioned in Section 13020 [see text] and from any other appropriate source.

(b) To prepare and distribute to all those persons and agencies cards, forms, or electronic means used in reporting data to the department. The cards, forms, or electronic means may, in addition to other items, include items of information needed by federal bureaus or departments engaged in the development of national and uniform criminal statistics.

(c) To recommend the form and content of records that must be kept by those persons and agencies in order to ensure the correct reporting of data to the department.

(d) To instruct those persons and agencies in the installation, maintenance, and use of those records and in the reporting of data therefrom to the department.

(e) To process, tabulate, analyze, and interpret the data collected from those persons and agencies.

(f) To supply, at their request, to federal bureaus or departments engaged in the collection of national criminal statistics data they need from this state.

(g) To make available to the public, through the department's OpenJustice Web portal, information relating to criminal statistics, to be updated at least once per year, and to present at other times as the Attorney General may approve reports on special aspects of criminal statistics. A sufficient number of copies of a downloadable summary of this information shall be annually prepared to enable the Attorney General to send a copy to the Governor and to all public officials in the state dealing with criminals and to distribute them generally in channels where they will add to the public enlightenment. This subdivision shall not be construed to require more frequent reporting by local agencies than what is required by any other law.

(h) To periodically review the requirements of units of government using criminal justice statistics, and to make recommendations for changes it deems necessary in the design of criminal justice statistics systems, including new techniques of collection and processing made possible by automation.

(i) To evaluate, on an annual basis, the progress of California's transition from summary crime reporting to incident-based crime reporting, in alignment with the federal National Incident-Based Reporting System, and report its findings to the Legislature annually through [2019].

SOURCE: Section 5 of Assembly Bill 2524, signed into law September 21, 2016.

such "data they need from this state" (Penal Code § 13010(f); this wording unchanged under the OpenJustice Data Act described below and in Box E.2).

- *State has no incident-based reporting system, and was for years listed among the states with no announced intent to transition to NIBRS—but that has changed recently:* Signed into law on September 21, 2016, Assembly Bill No. 2524 (the OpenJustice Data Act of 2016) builds off of the 2015 deployment of the state's OpenJustice web portal for dissemination and visualization of crime and justice data. The act includes among its legislative findings/declarations the following:
  - "The Department of Justice is working with the Federal Bureau of Investigation and the federal Bureau of Justice Statistics to transition California from summary crime reporting to incident-based crime reporting," and
  - "It is the intent of the Legislature that following full implementation of incident-based crime reporting, the Department of Justice work to transition to exclusively electronic crime data collection and evaluate the potential for criminal justice statistical data to be updated on the OpenJustice Web portal at a frequency greater than once per year."

  The act amends the core of the state's UCR law, Section 13010 of the Penal Code, to read as shown in Box E.2, substantially revising clause (g) and adding the new clause (i). The act does not commit to a timeline for NIBRS conversion but clarifies intent to move in that direction. To further the state's work, the California state program received an NCS-X Planning grant; the Long Beach, Los Angeles, and San Diego police departments and Los Angeles County sheriff's department have been awarded NCS-X development funds.
- *Annual crime report required by law:* Prior to the OpenJustice Data Act of 2016, the Attorney General was obligated to submit "an annual report containing the criminal statistics of the preceding calendar year" to the governor on or before July 1. Likewise, the prior text of Penal Code § 13010(g) required that all reports be printed in sufficient quantity that they may be sent to "all public officials" and "distribute[d] generally in channels where they will add to the public enlightenment." The new language, in Box E.2, signals a move to more frequent and continuous reporting online.
- *Other key features:* In 2014, California compiled a handbook for criminal justice and related agencies, clarifying the various requirements for crime reporting in the state and the due dates for submitted data (California Department of Justice, 2014). Among the entries in the handbook—and offense types—singled out for special collection in California are:
  - *Anti-reproductive-rights offenses,* identified in the handbook as "crimes that are committed against reproductive health services providers, clients, assistants, for the facilities where those services are provided or at a place of worship because of the church's beliefs regarding reproductive rights." These data are collected under authority of Penal Code § 13777(a) and are incident- and attribute-based in nature (including location, race and gender

of victims and suspects, and weapon involved), and are summarized in a separate annual report.
- *Citizens' complaints against peace officers,* involving both criminal and noncriminal complaints filed by citizens against law enforcement personnel. Penal Code § 13012(e) requires that the annual report on crime in California include these data; that said, the code mandates that "the report shall not contain a reference to any individual agency but shall be by gross numbers only."
- *Domestic violence calls for assistance,* regardless of whether the calls resulted in an arrest or further action, as required by Penal Code § 13730.
- *Homicide.* Akin to the Supplementary Homicide Report of the national UCR program, California requires the collection of additional demographic and contextual information on every homicide incident. Penal Code § 13014(b) requires every "state or local governmental entity" responsible for investigation of homicide to complete these reports, in parallel with usual crime reporting.

• *State Statistical Analysis Center:* Same entity, Criminal Justice Statistics Center in Department of Justice.

COLORADO

• *Local law enforcement agencies required to submit crime statistics to state entity:* The Colorado Bureau of Investigation is tasked by CO Rev Stat § 24-33.5-412(1) to, among other duties, "operate the statewide uniform crime reporting program." § 24-33.5-412(5) goes on to stipulate that "every law enforcement agency in this state shall furnish such information to the bureau concerning crimes, arrests, and stolen and recovered property as is necessary for uniform compilation of statewide reported crime, arrest, and recovered property statistics." Colorado is unusual for suggesting some limited cost sharing in the same passage: "The cost to the law enforcement agency of furnishing such information shall be reimbursed out of appropriations made therefor by the general assembly; except that the general assembly shall make no such reimbursement if said cost was incurred in a fiscal year during which the Colorado crime information center was funded exclusively by state or federal funds."

• *Transfer to FBI/national compilation not directly required by law,* though that is consistent with the requirement under the law (§ 24-33.5-412(5)) that obligates the bureau to adopt reporting forms and "reporting standards consistent with the development of the strategic plan for an integrated [state] criminal justice information system" that is, in turn, "consistent with applicable federal and state laws and regulations such as the national criminal justice information system standards."

- *State does not maintain incident-based reporting system but is NIBRS certified:* As per the JRSA's state profiles, Colorado was certified by the FBI to submit NIBRS data in 1997. At the time of the JRSA profile, NIBRS-reporting agencies covered 90 percent of the population of Colorado and accounted for 88 percent of the reported crime. The most recent *Crime in Colorado* report, for 2016, notes the number of agencies reporting data to the state (in either NIBRS or Summary format) as 244 but states only that "Colorado law enforcement agencies report incident-based statistics, which are collected and stored in the Colorado Crime Information Center (CCIC) computer located at the Colorado Bureau of Investigation" (http://crimeinco.cbi.state.co.us/cic2k16/intro.php). The FBI's 2015 NIBRS publication suggests that 193 Colorado agencies submitted 12 months of NIBRS data for the year.
- *Annual crime report not apparently referenced in law,* though implied—indeed, the computer crime passage described below requires "reports at least annually concerning activities of computer crime in Colorado."
- *Other key features:* The key enabling statute for the state's UCR program, § 24-33.5-412(5), singles out offenses of "murder, sexual assault, [robbery]," or "child abuse or sexual assault on a child" for special treatment; submitted information for such offenses is required to include "the modus operandi of such crimes in order to facilitate the identification of cross-jurisdictional offenders." § 24-33.5-412(8) authorizes the bureau to investigate computer crime violations but also to "develop and collect information with regard to computer crime in an effort to identify, charge, and prosecute criminal offenders and enterprises that unlawfully access and exploit computer systems and networks, impact functionality, and access sensitive data and shall report such information to the appropriate law enforcement organizations."
- *State Statistical Analysis Center:* Housed in state Division of Criminal Justice.

CONNECTICUT

- *Local law enforcement agencies required to submit crime statistics to state entity:* Connecticut's Department of Emergency Services and Public Protection is directed (CT Gen Stat § 29-1c(a)–(b)) to "establish a state-wide uniform crime reporting system," in which "each organized police department shall participate [by] submitting [at] such times and on such forms as [prescribed], a uniform crime report concerning crimes committed within such organized police department's jurisdiction. Such report shall contain the number and nature of offenses committed and such other information as the [Department] shall require."
- *Transfer to FBI/national compilation required by law:* A distinctive feature of Connecticut's law is that the FBI is designated as the arbiter of whether

a local agency's submissions are complete and correct. Per § 29-1c(c), the Department is obligated to notify the "chief elected official of the municipality" whose law enforcement agency either fails to report or provides "missing, incomplete or incorrect information"—and inform said official "of the consequences" of these problems, to include ineligibility for state or federal law enforcement grants. "For purposes of this subsection, information is missing, incomplete or incorrect if so designated by the Federal Bureau of Investigation or if not submitted within sixty days of the end of the month in which such information is required to be reported."

- *State does not have its own incident-based reporting system, but is certified for NIBRS reporting:* As of the JRSA inventory, 102 law enforcement agencies reported to the state UCR program, with NIBRS format representing 75 percent of the state's population and 59 percent of the observed crime. "By the end of 2015, 85 of 104 Connecticut law enforcement agencies were reporting NIBRS data to the statewide repository," the additional two agencies entering the Connecticut system being the Mashantucket and Mohegan Tribal Nations (Uniform Crime Reporting Program, 2016:9–10).
- *Annual crime report required by law:* The Department is required by law to "publish an annual report concerning the extent, fluctuation, distribution and nature of crime in Connecticut," with "specific analysis of the nature, extent and pattern of sex crimes in the state" (§ 29-1c(e)).
- *State Statistical Analysis Center:* Housed in state Office of Policy and Management.

DELAWARE

- *Local law enforcement agencies required to submit crime statistics to state entity:* The grounding of the Delaware UCR program in state law is somewhat unclear, inasmuch as it appears to derive solely from the text at 11 DE Code § 8901(b):

> The General Assembly finds that there is a need:
> (1) To designate the State Bureau of Identification [of the Delaware State Police] as the central state repository for criminal history record information;
> (2) To require the rapid identification, classification and filing of fingerprints;
> (3) To require the reporting of accurate, relevant and current information to the central repository by all criminal justice agencies;
> (4) To insure that criminal history record information is kept accurate and current; and
> (5) To prohibit the improper dissemination of such information.

- *Transfer to FBI/national compilation not apparently specified by law*, but done in practice.

- *State maintains own incident-based reporting system and is fully NIBRS compliant:* Per the JRSA profiles, nearly all Delaware agencies were certified for NIBRS in 2001, and the remainder followed in the next few years; it now lists as 100% covered by NIBRS-format, reported by 71 agencies. The methodology section of the most recent *Crime in Delaware* report indicates that the State Bureau of Identification Quality Control Section takes primary responsibility for working with raw data files derived from local agencies' incident reports—recoding them for compliance with NIBRS standards. (The Dover, New Castle County, and Wilmington Police Departments are sufficiently large and resourced that they are described as performing extensive data quality routines on their own, as well). It is the state-processed data that is transmitted to the FBI (Delaware Criminal Justice Council, Statistical Analysis Center, 2016).
- *Annual crime report not apparently specified by law.* In practice, Delaware's SAC (in collaboration with the State Bureau of Identification) issues an annual report that focuses on a 5-year window of data (e.g., the August 2016 report included 2011–2015 data).
- *State Statistical Analysis Center:* Vested by law in state Criminal Justice Council.

FLORIDA

- *Local law enforcement agencies not explicitly directed by law to submit crime statistics to state entity:* FL Stat § 943.05(2)(c) tasks the Criminal Justice Information Program within the Florida Department of Law Enforcement (FDLE) tasked to "initiate a crime information system that shall be responsible for: 1. Preparing and disseminating semiannual reports to the Governor, the Legislature, all criminal justice agencies, and, upon request, the public. Each report shall include, but not be limited to, types of crime reported, offenders, arrests, and victims. 2. Upon request, providing other states and federal criminal justice agencies with Florida crime data" Under § 943.05(2)(d), FDLE is granted rulemaking authority "to effectively and efficiently implement, administer, manage, maintain, and use the automated biometric information system and uniform offense reports and arrest reports. The rules shall be considered minimum requirements and shall not preclude a criminal justice agency from implementing its own enhancements."
- *Transfer to FBI/national compilation not apparently addressed by law.*
- *State does not maintain incident-based reporting system, and has long been listed among the states with no plans to transition to NIBRS.* According to a history page maintained by the FDLE (http://www.fdle.state.fl.us/cms/FSAC/Data-History.aspx), Florida is a very rare case in which detailed

incident-based reporting was tried and rejected in favor of summary-format reporting. The page indicates:

> In 1976, Florida was one of the first states to abandon the traditional method of collecting UCR data on a monthly basis and initiated a sophisticated offense-by-offense or incident-based reporting program. Municipal police and county sheriff's departments reported the individual offense data known to them in the following crime categories: murder, non-traffic manslaughter, forcible rape, robbery, assault, burglary, larceny, motor vehicle theft, and arson. In addition, local law enforcement agencies reported supplemental offense information such as the number of victims, victim's age, sex and relationship to offender, type of weapon used, the value of property stolen and recovered, circumstances surrounding homicides and other pertinent crime data.

The transition was apparently not smooth, with 1989 being the "first full year of data under the enhanced UCR program"—and that said to be after the program endured a rough "transition year" in 1988, when several reporting agencies were unable to accommodate revised system and data format requirements. Finally, "in 1996, the Florida UCR changed its data collection method from a monthly reporting of crime incidents to a semi-annual reporting of summary offense data. With the change from an incident based system to a summary based system, supplementary data such as victim information are no longer available on all offense data except homicide." Summary-only reporting from FDLE to the national UCR program has continued through the present, though the FDLE was a recipient of a National Crime Statistics Exchange (NCS-X) grant in 2015 to develop a plan for equipping a state-level, NIBRS-certified system and program.

- *Annual crime report required by law:* Indeed, the construction of Florida's law (quoted above) in § 943.05(2)(c) is unusual among states for listing reports to the governor and state officials (and, "upon request, the public") as the principal purpose for collection of crime information.
- *Other key features:* Separate units set up within the FDLE, tasked with information/database development in conduct of their duties, include the Financial Crime Analysis Center (targeted at "money laundering, drug-related criminal activity, or other criminal activity involving financial transactions"; § 943.032), the Domestic Security and Counter-Terrorism Intelligence Center (§ 943.0321), and the Cybercrime Office (§ 943.0415). Domestic violence is specifically called out as a priority for crime data collection and for annual reporting (§ 943.1702).
- *State Statistical Analysis Center:* Same entity, Florida Department of Law Enforcement.

GEORGIA

- *Local law enforcement agencies required to submit crime statistics to state entity:* The Georgia Crime Information Center in the Georgia Bureau of Investigation (O.C.G.A. § 35-3-33) is obligated to "provide a uniform crime reporting system for the periodic collection, analysis, and reporting of crimes reported to and otherwise processed by any and all law enforcement agencies within the state, as defined and provided for in this article." Participation in the UCR program is made binding by § 35-3-36 (2015): "All criminal justice agencies within the state shall submit to the center, periodically at a time and in such form as prescribed by the center, information regarding only the cases within its jurisdiction and in which it is or has been actively engaged. Such report shall be known as the 'uniform crime report' and shall contain crimes reported and otherwise processed during the period preceding the period of report, including the number and nature of offenses committed, the disposition of such offenses, and such other information as the center shall specify, relating to the method, frequency, cause, and prevention of crime."
- *Transfer to FBI/national compilation required by law:* Under § 35-3-33, the center is required to "cooperate with other agencies of the state, the crime information agencies of other states, and the Uniform Crime Reports and National Crime Information Center systems of the Federal Bureau of Investigation in developing and conducting an interstate, national, and international system of criminal identification, records, and statistics."
- *State has no incident-based reporting system but has recently committed to NIBRS conversion:* As of the JRSA profile, Georgia was listed as having "no plans" for NIBRS adoption, though one local agency (Norcross) had been certified by the FBI. However, in August 2015, the Georgia Bureau of Identification (GBI) publicly committed to converting to NIBRS ahead of the 2021 sunset of the UCR Summary Reporting system (and ideally by 2018). As described in an article for an FBI newsletter (at https://www.fbi.gov/services/cjis/cjis-link/going-nibrs-part-2-georgia-transition):

  > The GBI's monthly report for June 2016 declared the state's plan to effect a transition to NIBRS. The advanced notice allows these agencies time to plan for the associated costs of NIBRS transition. While the GBI will assume the costs associated with the implementation of the state data system, local law enforcement agencies could incur costs if they wish to integrate their local records management systems with the GBI's NIBRS-compliant system. In August 2016, the GBI issued a Request for Proposal to procure a service provider to implement a state NIBRS system.

- *Annual crime report required by law:* Part of the base statute authorizing the crime reporting program (§35-3-33) requires the center to "compile statistics on the nature and extent of crime in the state and compile other data related to planning for and operating criminal justice agencies, provided that such statistics do not identify persons, and make available all such statistical information obtained to the Governor, the General Assembly, and any other governmental agencies whose primary responsibilities include the planning, development, or execution of crime reduction programs. Access to such information by the latter governmental agencies will be on an individual, written request basis wherein must be demonstrated a need to know, the intent of any analyses, dissemination of such analyses, and any security provisions deemed necessary by the center."

- *Other key features:* Georgia law requires participating agencies to comply with periodic quality control audits (§ 35-3-36): "Upon the request of the center, local law enforcement agencies shall periodically provide for audit samples of incident reports for the preceding reporting period so that the center may help ensure agency compliance with national and state uniform crime reporting requirements."
- *State Statistical Analysis Center:* Part of Georgia Criminal Justice Coordinating Council.

HAWAII

- *Local law enforcement agencies required to submit crime statistics to state entity:* The Hawaii Criminal Justice Data Center is established under the auspices of the state attorney general by HI Rev Stat § 846-2, and is deemed the state UCR program by § 846-2.5: Through the center, "the attorney general shall provide for the collection, recording, and compilation of data and statistics relating to crime. ... The systems [of criminal identification and crime statistics] shall be uniform throughout the State, shall be continuous in operation, and shall be maintained as far as possible in a manner as shall be in keeping with the most approved and modern methods of identification and of the collection and compilation of the statistics." Reporting of offense data is made mandatory by § 846-18's provision that "the chiefs of police of the counties and state and county agencies having the power of arrest shall provide to the attorney general crime incident reports and any other information regarding crimes committed within their respective jurisdictions as necessary for the attorney general to operate a statewide crime reporting program and to cooperate with the Federal Bureau of Investigation's uniform crime reporting program. The reports and information shall be submitted to the attorney general in the manner, form, and time schedule as the attorney general may prescribe."
- *Transfer to FBI/national compilation required by law:* As quoted above, it is deemed necessary for the attorney general to cooperate with the national UCR program (§ 846-18).
- *State has no incident-based reporting system,* but was awarded NCS-X implementation grant funding in 2016.

- *Crime report required by law,* albeit not on fixed schedule. To fulfill its purpose, the data center is required to "develop systems and provide the structure that support criminal justice information systems, provide statistical research and data analysis, and make public periodic reports which shall provide the public with a clear view of the criminal justice systems" (§ 846-2.5).
- *Other key features:* § 846-2.5 is unusually specific relative to other state codes about the funding responsibilities between state and local governments: "The several counties shall provide the necessary equipment and the compensation of the persons required to install and carry out the work of the systems of identification and statistics in their respective jurisdictions; provided that those expenses in connection with matters exclusively within the control of the State shall be borne by the State" (to the latter point, the state bears the burden for the state's fingerprint system).
- *State Statistical Analysis Center:* Same department, part of Hawaii Department of the Attorney General.

IDAHO

- *Local law enforcement agencies required to submit crime statistics to state entity:* ID Code § 67-3003 directs the Bureau of Criminal Identification (of the Idaho State Police) to "establish a uniform crime reporting system for the periodic collection and reporting of crimes, and compile and publish statistics and other information on the nature and extent of crime in the state." Per § 67-3006, "a law enforcement agency shall submit to the department, at the time, in the manner, and in the form prescribed by the department, data regarding crimes committed within that agency's jurisdiction."
- *Transfer to FBI/national compilation encouraged by law:* The state code hews to the construct of cooperation with other entities in outlining a general duty of the Bureau of Criminal Identification: "Cooperate with other criminal justice agencies of the state, state and federal courts, the criminal records repositories of other states, the federal bureau of investigation criminal justice information services, the national law enforcement telecommunications system, and other appropriate agencies and systems, in the operation of an effective interstate and national system of criminal identification, records and statistics" (§ 67-3003).
- *State is fully NIBRS compliant:* The JRSA inventory notes that Idaho was certified to report NIBRS in 1992; introductory text in the most recent *Crime in Idaho* report suggests that NIBRS-format collection was initiated in 1991 and has continued, by all agencies within the state, since that time (Idaho State Police, 2017:iv). Local departments in the state are offered three options for submitting their NIBRS-format data: direct online entry using a

state law enforcement telecommunications system, submission of electronic data files from approved records management systems, or hard-copy paper form ("which is limited to agencies with less than 50 reportable incidents a month") (Idaho State Police, 2017:iv).

- *Annual crime report required by law:* Reporting/publication is listed as the primary purpose of crime data collection in § 67-3003, quoted above. The reporting function is reinforced in § 67-3006, which requires that "the department shall publish an annual report, available no later than July 1 of the following year, containing the statistical information gathered under this section that relates to the number and nature of criminal offenses, arrests, and clearances, and any other data the director determines to be appropriate relating to the method, frequency, cause and prevention of crime."
- *Other key features:* Idaho is one of several states that defines an offense of "malicious harassment"—essentially, bias-motivated assault or threat, or hate crimes. § 67-2915 of the state code obligates the director of the Idaho State Police to "annually submit" to the governor and state legislative leaders "a report on malicious harassment crimes," in which "report content shall be limited to statistical data." To facilitate this, malicious harassment reporting is required of local departments in parallel with standard UCR reporting: "All city, county and state law enforcement units shall be required to report to the director all incidences of, complaints on, and arrests for malicious harassment crimes within their respective jurisdictions. The director shall develop a standard procedure and shall prescribe and provide a standard form for complete and uniform reporting."
- *State Statistical Analysis Center:* Same department (Idaho State Police), though placed in Planning, Grants, and Research Department.

ILLINOIS

- *Local law enforcement agencies required to submit crime statistics to state entity:* In August 2015, Illinois Public Act 99-352 became law—a broader Police and Community Relations Improvement Act that included a new Uniform Crime Reporting Act as a section. Prior to this act (which became effective on January 1, 2016), the UCR program drew its mandate from a provision directing the Department of State Police to "investigate the origins, activities, personnel and incidents of crime and the ways and means to redress the victims of crimes, and study the impact, if any, of legislation relative to the effusion of crime and growing crime rates, and enforce the criminal laws of this State related thereto" (then, 20 ILCS 2605/2605-35). The new act retained language designating the State Police as "central repository and custodian of crime statistics for the State" (codified at 50 ILCS 709/5-10). Reporting is made mandatory of all local agencies: The State Police "shall have all the power necessary to carry out the purposes of this Act, including the power to demand and receive cooperation in the submission of crime statistics from all law enforcement agencies. All data and information provided to the Department under this Act must be provided in a manner and form prescribed by the Department" (50 ILCS 709/5-10).

- *Transfer to FBI/national compilation not apparently addressed by law*, though certainly implied (at least in part) by language quoted below obliging the State Police to require quarterly reports of arrest-related deaths "contingent upon updated federal guidelines regarding the Uniform Crime Reporting Program."
- *State currently has no incident-based reporting system, but is developing one:* In the FBI's 2015 NIBRS tabulations, as had been the case for many years, only one law enforcement in Illinois is NIBRS certified (and submitting)—the Rockford Police Department. However, numerous recent events portend change: the new UCR law, which requires attribute-intensive, incident-based data for some specific offense types (as described below); modernization of Chicago Police Department systems to facilitate the city's Open Data Portal and the department's CLEARMAP initiatives (and award of NCS-X pilot funding to Chicago); and award of NCS-X planning grant funds to the Illinois UCR Program.
- *Annual crime report required by law:* "On an annual basis, the Department shall make available compilations of crime statistics required to be reported by each law enforcement agency" (50 ILCS 709/5-10).
- *Other key features:* Perhaps the most interesting aspect of the construction of Illinois's new UCR law is that it simultaneously hints at a desire for incident-based (or at least attribute-heavy) data on some specific offense types, yet is still firmly committed to coarse summary/count data on *most* offense types. Specifically, the act (50 ILCS 709/512) requires:

> All law enforcement agencies shall submit to the Department of State Police on a monthly basis the following:
>
> (1) beginning January 1, 2016, a report on any arrest-related death that shall include information regarding the deceased, the officer, any weapon used by the officer or the deceased, and the circumstances of the incident. The Department shall submit on a quarterly basis all information collected under this paragraph (1) to the Illinois Criminal Justice Information Authority, contingent upon updated federal guidelines regarding the Uniform Crime Reporting Program;
>
> (2) beginning January 1, 2017, a report on any instance when a law enforcement officer discharges his or her firearm causing a non-fatal injury to a person, during the performance of his or her official duties or in the line of duty;
>
> (3) a report of incident-based information on hate crimes including information describing the offense, location of the offense, type of

victim, offender, and bias motivation. If no hate crime incidents occurred during a reporting month, the law enforcement agency must submit a no incident record, as required by the Department;

(4) a report on any incident of an alleged commission of a domestic crime, that shall include information regarding the victim, offender, date and time of the incident, any injury inflicted, any weapons involved in the commission of the offense, and the relationship between the victim and the offender;

(5) data on an index of offenses selected by the Department based on the seriousness of the offense, frequency of occurrence of the offense, and likelihood of being reported to law enforcement. The data shall include the number of index crime offenses committed and number of associated arrests; and

(6) data on offenses and incidents reported by schools to local law enforcement. The data shall include offenses defined as an attack against school personnel, intimidation offenses, drug incidents, and incidents involving weapons.

The act further obliges all law enforcement departments to "submit to the Department incident-based information on any criminal homicide" beginning July 1, 2016—akin to the national Supplementary Homicide Reports. The operative clause in this passage—designated fifth in a 6-item list—is the one referring to index crimes. Historically, the set of index crimes has been parsed as equivalent to the national UCR's Part I crimes (also called "index crimes," though the index is no longer calculated nationally)—except that Illinois only undertook a major "transition" in 2010, revisiting its definitions of index crimes to comport with national UCR standards.[1] For example, the state UCR program began to disallow designations of "aggravated assault" in cases where the assault victim is of a certain type (e.g., a police officer or a pregnant woman) even though Illinois statute classifies those as "aggravating" factors (Illinois State Police, 2016:3,6). In the 2015 law, it is striking that considerable detail is required for some offense types while the mandatory information on "index" crimes remains "the number of index crime offenses committed" and count of arrests.

The new UCR act in Illinois is also specific on the penalty for nonresponse by a local agency (50 ILCS 709/5-20):

> The Department of State Police shall annually report to the Illinois Law Enforcement Training Standards Board any law enforcement agency not in compliance with the reporting requirements under this Act. A law enforcement agency's compliance with the reporting requirements under this Act shall be a factor considered by the Illinois Law Enforcement

[1]Illinois also denotes violations of five specific state drug laws as index offenses, but only arrest information is collected for those crimes.

Training Standards Board in awarding grant funding under the Law Enforcement Camera Grant Act.

- *State Statistical Analysis Center:* Not colocated; independent Illinois Criminal Justice Information Authority.

INDIANA

- *Local law enforcement agencies required to submit crime statistics to state entity:* Indiana Code § 10-13-2-4 creates a Criminal Justice Data Division within the Indiana State Police to fulfill several purposes, two of which are to "inform the public and responsible governmental officials as to the nature of the crime problem, its magnitude, and its trend over time," and to "find out who commits crimes by age, sex, family status, income, ethnic and residential background, and other social attributes, [in order] to find the proper focus of crime prevention programs." Accordingly, it is empowered (§ 10-13-2-5) to "prepare and distribute to all the persons and agencies the forms to be used in reporting data to the division. The forms also must provide for items of information needed by federal bureaus, agencies, or departments engaged in the development of national criminal statistics." Local agencies and personnel, "if requested by the division," are required to "report to the division, as and when prescribed, all data requested" (§ 10-13-2-6).
- *Transfer to FBI/national compilation required by law:* As quoted above, the standard forms developed by the state are meant to include any data item requested by the national crime statistics program (and this is strengthened by text to take effect in 2018, discussed below). And the division's list of duties includes an item that obliges it to "supply data, upon request, to federal bureaus, agencies, or departments engaged in collecting and analyzing national criminal statistics" (§ 10-13-2-5).
- *State has a legally set deadline for conversion to incident-based reporting:* On July 1, 2018, revised text of § 10-13-2-6(a) will take hold, adding a step to the requirements of any local agency or official from whom data is requested by the state UCR program: "Report to the division all uniform crime data from the federal Uniform Crime Reporting Program using the National Incident Based Reporting System (NIBRS) format as required by the division." In the FBI's compilation of 2015 NIBRS data, the Indiana State Police reported 12 months of data for each of Indiana's 92 counties, but no municipal or other law enforcement agencies in the state reported.
- *Annual crime report required by law:* Another change in the enabling legislation will take effect on July 1, 2018. At present, the division is required to report crime statistics to state officials twice a year; the new text in 2018 will reduce that frequency to annual reporting. A second clause in that passage authorizes the division to report to the director "at other times

the superintendent considers necessary or the governor requests, reports on public aspects of criminal statistics in a sufficiently general distribution for public enlightenment" (§ 10-13-2-5).

- *Other key features:* Indiana's crime reporting law includes a distinctive appeal for consistent technical systems within the state—and broader—in addition to consistency in content. § 10-13-2-7 reads: "As far as is practicable, the equipment methods and systems used by the criminal justice data division must be compatible with those used by similar agencies in other states and the federal government so that data necessary for interstate, national, and international criminal justice is readily available." Indiana's code is also unusual in that it effectively immunizes local agency personnel from personal liability if the required data submission somehow causes harm: "An official required under this chapter to furnish reports, information, or statistics to the criminal justice data division or a person employed by the official is not liable in any action arising out of having furnished the information in a manner as may be required by this chapter or the rules adopted under this chapter" (§ 10-13-2-6(b)). In terms of penalties for nonresponse by local agencies or officials, Indiana code puts forth a somewhat vague penalty: In such cases, "the director of the criminal justice planning agency may deny the public official the benefits of the [criminal justice] system until the public official complies with the rules." However, the law is more decisive in addressing "an official who knowingly, intentionally, or recklessly makes a false return of information to the division"—that is deemed a Class A misdemeanor (§ 10-13-2-12).
- *State Statistical Analysis Center:* Not colocated; vested in separate Indiana Criminal Justice Institute.

IOWA

- *Local law enforcement agencies required to submit crime statistics to state entity:* The enabling statute for Iowa's UCR program, administered by the Department of Public Safety, has a distinctive construction relative to other states. As per IA Code § 692.15:

> If it comes to the attention of a sheriff, police department, or other law enforcement agency that a public offense or delinquent act has been committed in its jurisdiction, the law enforcement agency shall report information concerning the public offense or delinquent act to the department on a form to be furnished by the department not more than thirty-five days from the time the public offense or delinquent act first comes to the attention of the law enforcement agency. The reports shall be used to generate crime statistics.

- *Transfer to FBI/national compilation not apparently referenced by law.*

- *State maintains own incident-based reporting system:* Iowa's Incident-Based Uniform Crime Reporting program has been in operation since 1991, with most reporting agencies being certified to provide NIBRS format data circa 1992 and all agencies now reporting NIBRS format.
- *Annual crime report required by law:* "The department shall submit statistics to the governor, the general assembly, and the division of criminal and juvenile justice planning of the department of human rights on a quarterly and yearly basis" (IA Code § 692.15).
- *Other key features:* Parallel authority has been added to mandatory crime reporting of at least two offenses or crime classes—inserted into the state code at assorted places but following the same basic format. § 236.9 mandates that agencies "collect and maintain information on incidents involving domestic abuse" and report the same to the Department of Public Safety; the department, in turn, is required to issue reports "provided [that] individual identifying details of the domestic abuse are deleted." The clause also provides for the department to provide deidentified data "on individual incidents of domestic abuse to persons conducting bona fide research, including but not limited to personnel of the department of justice. The identically constructed § 692.22 requires the same data collection and reporting on the offense of stalking.
- *State Statistical Analysis Center:* Division of Criminal and Juvenile Justice Planning, Iowa Department of Human Rights.

KANSAS

- *Local law enforcement agencies required to submit crime statistics to state entity:* Two passages of state code define Kansas's incident-based UCR program. KS State § 21-2501a requires that "all law enforcement agencies ... shall maintain a permanent record of all felony and misdemeanor offenses reported or known to have been committed" within their jurisdictions; moreover, said record is to be kept "on forms approved by the attorney general" and a report on each such offense is supposed to be filed with the Kansas Bureau of Investigation (the manager of the UCR program) "within 72 hours after such offense is reported or known to have been committed." A simpler passage at § 21-2504 evokes the authorization for the national UCR program in U.S. Code, stipulating that, "for the purpose of controlling crime and obtaining reliable statistics about crime and criminals, the attorney general may call upon and obtain from the clerks of district courts, sheriffs, police departments and county attorneys all information that said attorney general may deem necessary in ascertaining the true condition of the crime situation; and it shall be the duty of the above-mentioned officers to furnish the information so requested by the attorney general."

- *Transfer to FBI/national compilation not apparently referenced by law.*
- *State maintains own incident-based reporting system:* The Kansas Incident-Based Reporting System was implemented in 1986 (by way of the enabling legislation described above) and was substantially revised in 1993 to be consistent with NIBRS requirements (Kansas Bureau of Investigation, 2012:5).
- *Annual crime report not apparently referenced by law.*
- *Other key features:* Parallel authority in state code (§ 22-2307) mandates that any domestic violence call result in the completion of a "standard offense report," to be filed to the Kansas Bureau of Investigation "whether an arrest is made or not." Kansas code is also unique because of language added to the UCR-enabling law in 2006: "It is hereby made the duty of every sheriff, police department or countywide law enforcement agency in the state to report within 30 days, on forms approved by the attorney general, any methamphetamine laboratory seizures or dump sites and any theft or attempted theft of anhydrous ammonia that occurs in such agency's jurisdiction" (KS Stat § 21-2501a).
- *State Statistical Analysis Center:* Separately vested in Kansas Sentencing Commission.

KENTUCKY

- *Local law enforcement agencies required to submit crime statistics to state entity:* Kentucky's UCR program derives its authority from KY Rev Stat § 16.060 and one of its stated duties assigned to the Kentucky State Police, wording that is reminiscent of the general charter in federal law given to the U.S. Department of Justice: The State Police are to "collect, classify and maintain information useful for the detection of crime and the identification, apprehension and conviction of criminals." To this end, § 16.130 mandates that "all peace officers of the various political subdivisions of this Commonwealth shall cooperate with the department and the Commissioner."
- *Transfer to FBI/national compilation required by law:* "The Commissioner shall cooperate and exchange information with any other department or agency of this Commonwealth, or with other police forces, both within and without this Commonwealth and with the federal government" (§ 16.130).
- *State maintains own incident-based reporting system and is considered completely NIBRS compliant:* Kentucky switched to requiring incident-based reporting on July 1, 1987. "The state has been certified for [NIBRS] reporting, which is already taking place in the majority of law enforcement agencies within Kentucky" (Kentucky State Police, 2014:2)
- *Annual crime report required by law:* As per § 17.153, "The annual report of the department shall contain statistics showing: (a) The number and

type of offenses known to public authorities; (b) The personal and social characteristics of criminals and delinquents; and (c) The administrative action taken by law enforcement, judicial, penal, and correctional agencies in dealing with criminals and delinquents." The law further directs that the report "interpret these statistics" and "and include statistics that are comparable with national criminal statistics published by federal agencies."

- *Other key features:* A clause that became effective in 2006 appears to mandate separate collection of information on both elder and child abuse, as well as incidences of domestic violence, in parallel to regular UCR reporting. As per § 15A.190:

  > The Justice and Public Safety Cabinet [the administrative parent of the Kentucky State Police], in consultation with the Cabinet for Health and Family Services, the Kentucky Commission on Women, and any other agency concerned with particular acts of criminal activity, shall design, print, and distribute to all law enforcement agencies in the Commonwealth, a uniform reporting form which provides statistical information relating to the crimes involving domestic violence, child abuse, victimization of the elderly, including but not limited to elder abuse, neglect, and exploitation and other crimes against the elderly, or any other particular area of criminal activity deemed by the secretary of justice and public safety to require research as to its frequency.

- *State Statistical Analysis Center:* Also vested in Justice and Public Safety Cabinet, the administrative home of the Kentucky State Police.

LOUISIANA

- *Local law enforcement agencies required to submit crime statistics to state entity:* LA Rev Stat § 15:578 vests the Louisiana Bureau of Criminal Identification and Information with broad authority to "establish, maintain, and regulate a modern system of telecommunication and data processing for the efficient collection, storage, and rapid transmission of criminal history record information and relevant statistics." However, clause D of the same section formally transfers "the functions, powers, and duties of the bureau related to the gathering and dissemination of data concerning parish prison population statistics and uniform crime reports" to the state's separate Commission on Law Enforcement and the Administration of Criminal Justice, the primary duty of which is to coordinate the activities of the state's correctional systems. Under § 15:1204.3, legislative language that dates to 1991, "all law enforcement agencies, correctional agencies and institutions, district attorneys and municipal prosecutors, courts having criminal jurisdiction, or any other public agency dealing with crimes or criminals" are mandated to "install and maintain records needed for reporting crime incident and arrest information" and to report the same "to the commission as and when the commission prescribes." The commission "may impose reasonable administrative sanctions as it deems appropriate" for nonresponse to UCR reporting requirements, "includ[ing] the preclusion of a subject agency's participation in any of the grant programs operated by the commission" (§ 15:1204.5)

- *Transfer to FBI/national compilation required by law:* The content of the data collection instruments in the Louisiana system "shall also provide for other items of information needed by federal and state bureaus or departments engaged in the development of national and state statistics," and it is a duty of the state UCR program to "supply data to federal bureaus or departments engaged in collecting national criminal statistics" (§ 15:1204.5B(4),(7)).
- *State maintains own incident-based reporting system:* The Louisiana Incident-Based Reporting System was certified by the FBI as a source for NIBRS data in 2003; it was envisioned as an eventual replacement for Summary-format reporting but made available to local agencies as a reporting option on a voluntary basis. "By December 2010, there were 40 certified LIBRS agencies in Louisiana" (Louisiana Uniform Crime Reporting Program, 2016:7); the FBI's 2015 NIBRS compilation includes 12 months of NIBRS data from 25 Louisiana agencies. Incident-based reporting is directly written into the state's UCR law; one of the program's duties is "to instruct [persons] and agencies in the installation, maintenance, and use of incident-based records and in the manner of reporting to the system" (§ 15:1204.5B(5)). The state program was awarded an NCS-X Implementation grant to facilitate expansion of the system to more agency contributors.
- *Annual crime report required by law:* The commission is required "to present annually to the governor and members of the legislature, on or before May first, a report containing the criminal statistics of the most recently available calendar year and to present at such other times as the commission may deem wise, or [as requested by the governor or state legislative leaders], reports on special aspects of criminal statistics" (§ 15:1204.5(8)(a)).
- *Other key features:* Access to crime statistics data is given unusual weight in Louisiana law, put forward as the first-listed function/power of the state UCR program: "To establish, through electronic data processing and related procedures, a system by which relevant information can be collected, coordinated, analyzed, and made readily available to serve qualified agencies concerned with the administration of criminal justice located anywhere in the state. The commission shall prescribe the terms and conditions under which such agencies shall contribute or gain access to information contained in the system files" (§ 15:1204.2B(1)). Local law enforcement agencies are obligated by the law to "give the commission access to closed criminal case records for purposes of inspection," likely for quality control/audit purposes (§ 15:1204.3) A special requirement of Louisiana law is that the annual report of crime statistics must include "a category entitled 'gaming related

offenses' and shall include statistics involving violations of [various penal code provisions] in which the offender's gaming activity is a motive for commission of the crime" (§ 15:1204.2B(8)(b)).

- *State Statistical Analysis Center:* Same entity, Louisiana Commission on Law Enforcement.

MAINE

- *Local law enforcement agencies required to submit crime statistics to state entity:* Maine's UCR program is established within the State Bureau of Identification by 25 ME Rev Stat § 1544:

> It shall be the duty of all state, county and municipal law enforcement agencies, including those employees of the University of Maine System appointed to act as policemen, to submit to the State Bureau of Identification uniform crime reports, to include such information as is necessary to establish a Criminal Justice Information System and to enable the commanding officer to comply with section 1541, subsection 3 [cooperate 'with similar bureaus in other states and with the national bureau in the Department of Justice in Washington, D.C.']. It shall be the duty of the bureau to prescribe the form, general content, time and manner of submission of such uniform crime reports. The bureau shall correlate the reports submitted to it and shall compile and submit to the Governor and Legislature annual reports based on such reports. A copy of such annual reports shall be furnished to all law enforcement agencies.

- *Transfer to FBI/national compilation required by law,* as quoted above.
- *State is developing/testing incident-based reporting system:* As of the JRSA inventory, agencies covering 23 percent of Maine's population and 25 percent of its crime reported data in NIBRS format, and the state program was FBI certified. The state began work on a new state-wide information management system, the Maine Criminal Justice Information System, in 2015, which included an assessment of high-level requirements by the state UCR program. Later that year, "Maine identified a software vendor capable of delivering a NIBRS-compliant repository package" and has since conducted testing with "prototype police field report forms" (Maine State Police, 2016:13). Maine has been awarded an NCS-X for implementation/expansion.
- *Annual crime report not apparently referenced in law.*
- *Other key features:* Maine singles out animal cruelty as a priority crime-reporting category, along with bias-motivated crimes: "The bureau shall establish a category for abuse by adults of family or household members, a category for cruelty to animals and a category for crimes that manifest evidence of prejudice based on race, religion, disability, sexual orientation or ethnicity that are supplementary to its other reported information. The bureau shall prescribe the information to be submitted in the same manner

as for all other categories of the uniform crime reports" (25 ME Rev Stat § 1544). Separate statute enacted in 1979 mandates collection of data on domestic violence incidents, or "incidents of abuse by adults of family or household members" (recodified in 1995 in 19-A ME Rev Stat § 4012).

- *State Statistical Analysis Center:* Muskie School of Public Service, University of Southern Maine.

MARYLAND

- *Local law enforcement agencies required to submit crime statistics to state entity:* Maryland's Secretary of Public Safety is empowered to "establish and modify systems for receiving, processing, and maintaining [reports] and records of occurrences or alleged occurrences of crime and motor vehicle accidents in the State" (MD Pub Safety Code § 2-204). There is no apparent legal mandate for local agencies to submit *all* crime statistics to the Department of State Police, to whom the authority is delegated—the apparently operative passage of the code stresses cooperation, that the State Police's powers and duties "with respect to law enforcement are supplemental to and concurrent with similar powers and duties conferred by law on other law enforcement agencies of the State in their respective jurisdiction. ... The duties imposed on the Department [shall] promote effective cooperation among law enforcement agencies [and] do not limit the powers or responsibilities of any other law enforcement agency" (§ 2-302). However, a general obligation to report crime data to the state can be inferred from § 2-307, which does explicitly direct "each local law enforcement agency and the State Fire Marshal" to provide data on bias-motivated crimes.
- *Transfer to FBI/national compilation not apparently referenced by law.*
- *State does not maintain incident-based reporting system:* The state UCR program has been awarded an NCS-X grant for NIBRS-compliant system planning purposes, and the police departments of Montgomery County and Prince George's County have both been awarded NCS-X funds.
- *Annual crime report required by law,* albeit not necessarily on an annual cycle. § 2-307 assigns authority, "in general," to the State Police to "collect, analyze, and disseminate information about the incidence of crime in the State"—but, specific to hate crimes, to "make monthly reports to the Commission on Civil Rights." A subsequent passage (§ 2-308) obliges the State Police to "periodically publish statistics on the incidence of crime in the State" and "at least monthly [to] publish statistics about the occurrence and cause of all motor vehicle accidents in the State."
- *State Statistical Analysis Center:* Governor's Office of Crime Control and Prevention.

MASSACHUSETTS

- *Local law enforcement agencies not apparently explicitly mandated to submit crime statistics to state entity:* The General Laws of Massachusetts are intricate, but seem to share with Maryland the unusual structure that more is defined by law with specific regard to hate crimes (and other specific offense types) than in general. Massachusetts's UCR program is the Crime Reporting Unit of the State Police, part of the Criminal Information Section, Division of Investigative Service; the State Police department, in turn, is part of the Executive Office of Public Safety and Security. The Crime Reporting Unit is—quite literally—defined in a passage of law setting out a glossary for subsequent sections: It is "a joint project of the department of state police and the department of criminal justice information services charged with the responsibility of collecting incident reports submitted by state, local and campus police departments and other law enforcement authorities and disseminating periodic reports analyzing and interpreting crime rates and trends in the commonwealth" (MA Gen L Chap. 22C § 32; lower case in original). Even specific to hate crime, the State Police are directed to promulgate regulations specifying "procedures for the solicitation and acceptance of reports regarding hate crimes which are submitted to the crime reporting unit"—consistent with purely voluntary contribution of data (MA Gen L Chap. 22C § 33).
- *Transfer to FBI/national compilation not apparently referenced by law.*
- *State was early adopter of incident-based reporting, on voluntary basis:* As per a descriptive page on the agency's website (http://www.mass.gov/eopss/law-enforce-and-cj/crime-stats-rsrch/), Massachusetts began collecting NIBRS-format data from participating agencies in the mid-1990s, winning certification by the FBI in 1995. The JRSA inventory suggests that NIBRS-format contributors account for 86 percent of the state's population and 82 percent of its crime; the CRU website suggests that "approximately one half of the full-time police agencies in the Commonwealth" report in NIBRS format. The reason for those figures being as far off from 100 percent as they are is because of the longstanding, traditional exception from NIBRS format reporting in Massachusetts: namely, the Boston Police Department. That said, the NCS-X initiative portends to remedy this gap, with Boston being awarded NCS-X development funds in 2016 to facilitate planning for NIBRS compliance.
- *Annual crime report required by law:* "Periodic reports" are required, by the definition quoted above (MA Gen L Chap. 22C § 32).
- *State Statistical Analysis Center:* Same entity, Executive Office of Public Safety and Security, albeit housed in Office of Grants and Research.

MICHIGAN

- *Local law enforcement agencies required to submit crime statistics to state entity:* Michigan's UCR law requires (MI Comp L § 28.251, originally enacted 1968) "the police department of each city or village, any duly constituted police department of a township, and the sheriff's department of each county" to submit a crime report "once each month" to the Department of State Police. Said report "shall cover crimes reported and otherwise processed" during the preceding month. In addition, the monthly report "shall contain the number and nature of offenses committed, the disposition of such offenses and such other information as the director of state police shall specify relating to the method, frequency, cause and prevention of crime," subject only to the restriction that "under no circumstances shall the name of any person be reported." § 28.253 requires the State Police to include in their compilation any information voluntarily submitted by "any governmental police agency, not falling within the description of those required to submit the monthly uniform crime report."
- *Transfer to FBI/national compilation required by law:* The State Police are obliged (by § 28.252) to "prepare a statewide compilation of the statistics contained therein" and to make that compilation "available to any governmental law enforcement agency in the state, the judiciary committees of the Michigan state senate and the Michigan state house of representatives, and the federal bureau of investigation, upon request."
- *State maintains own incident-based reporting system,* dubbed MICR (Michigan Incident Crime Reporting), and is fully NIBRS compliant. Many Michigan law enforcement agencies have been NIBRS certified since the late 1990s; as of 2005, the state UCR program required to all agencies to report in MICR format.
- *Annual crime report required by law,* specifically the passage quoted above (§ 28.252) under transfer to national program.
- *Other key features:* Separate provisions in the state's UCR law require detailed data collection on domestic violence incidents and on hate crimes. With respect to the former, all police chiefs and sheriffs are obligated to report (by § 28.257):

> the following information related to domestic violence incidents:
> (a) The number of assaults reported that involve an adult and a minor and the disposition of those offenses.
> (b) The number of assaults reported that involve either 2 male adults or 2 female adults and the disposition of those offenses.
> (c) The number of assaults reported that involve 1 male adult and 1 female adult and the disposition of those offenses.
> (d) The number of crimes reported that involve an individual and his or her spouse, his or her former spouse, an individual with whom he

> or she has had a child in common, an individual with whom he or she has or has had a dating relationship, or an individual who resides or has resided in the same household; and the disposition of those offenses. . . .
>
> (e) Other statistics the director of the department of state police considers necessary to obtain accurate and reliable data on the incidence of domestic violence in this state.

Hate crime reporting is mandated by § 28.257a.

- *State Statistical Analysis Center:* Michigan Justice Statistics Center, Michigan State University.

MINNESOTA

- *Local law enforcement agencies required to submit crime statistics to state entity:* Minnesota's UCR program is housed in the Bureau of Criminal Apprehension in the Department of Public Safety. Its basis in law stems from text originally dating to 1927, in MN Stat § 299C.05:

> It shall be the duty of this division to collect, and preserve as a record of the bureau, information concerning the number and nature of offenses known to have been committed in the state, of the legal steps taken in connection therewith from the inception of the complaint to the final discharge of the defendant, and such other information as may be useful in the study of crime and the administration of justice. The information so collected and preserved shall include such data as may be requested by the United States Department of Justice, at Washington, under its national system of crime reporting. To the extent possible, the superintendent must utilize a nationally recognized system or standard approved by the Federal Bureau of Investigation to collect and preserve crime data.

§ 299C.06 makes it "the duty of all sheriffs, chiefs of police, prison wardens, superintendents of hospitals for persons with mental illnesses, reformatories and correctional schools, probation and parole officers, school attendance officers, coroners, county attorneys, court clerks, the commissioner of public safety, the commissioner of transportation, and the state fire marshal to furnish to the division statistics and information regarding the number of crimes reported and discovered." Unless otherwise authorized, such submissions "must utilize a nationally recognized system or standard approved by the Federal Bureau of Investigation for reporting statistics and information."

- *Transfer to FBI/national compilation required by law,* as specified in passages quoted above.
- *State is in process of developing its own incident-based reporting system:* The state of Minnesota began the process of replacing its early-1970s vintage Criminal Justice Reporting System in 2013 with the assignment of dedicated state funds, having concluded that the current system "is built on obsolete

technology which is difficult to integrate with other systems and cannot be modified to meet changing state and federal reporting requirements." In fall 2014, conversion to NIBRS standards was integrated into the project plans, and three local departments were named as pilot agencies in June 2015 to begin NIBRS-format submission by early 2017. In October 2016, the Minnesota program was awarded an NCS-X Implementation award, to be used to "onboard" agencies sought for inclusion in the NCS-X sample of NIBRS-format providers.

Legislation originally enacted in 2005, codified as § 299C.40, establishes a Comprehensive Incident-Based Reporting System (CIBRS), also to be managed by the Bureau of Criminal Apprehension, but that system appears to be a modern repository of criminal history record information. To wit, the purposes of CIBRS spelled out in the law are all law-enforcement tactical in nature, such as to "prepare a case against a person [for] the commission of a crime]" or "conduct background investigations."

- *Annual crime report not apparently required by law.*
- *Other key features:* Two unique features of Minnesota's crime reporting laws include:
  - *Explicit prohibition of considering suicide to be a crime:* Pursuant to § 299A.76, the Commissioner of Public Safety must not "include any statistics on committing suicide or attempting suicide in any compilation of crime statistics," nor "label as a crime statistic, any data on committing suicide or attempting suicide." The offenses of aiding suicide (and aiding attempted suicide) are not banned.
  - *Detailed study of human trafficking:* § 299A.785 authorizes the UCR program to "elicit the cooperation and assistance of government agencies and nongovernmental organizations as appropriate to assist in the collection of trafficking data" (stopping well short of mandating submission of data on trafficking-related offenses). The specific information sought on trafficking offenses includes "statistics on the number of trafficking victims, including demographics, method of recruitment, and method of discovery" and "social factors that contribute to and foster trafficking, especially trafficking of women and children." The Commissioner of Public Safety is required to compile and publish statistical data every 2 years.
- *State Statistical Analysis Center:* Minnesota Office of Justice Programs.

MISSISSIPPI

Among the states, Mississippi is currently the lone, major anomaly in terms of crime statistics reporting: It does not have a recognized state UCR program, and its state code appears to be completely silent on crime reporting.

News accounts have described spottiness of even Summary-format reporting; for instance, Apel (2015) observed that three of the state's 10 largest cities—Gulfport, Southaven, and Tupelo—do not appear in FBI UCR tabulations (Gulfport and Southaven are the second and third most populous cities in the state, respectively). The reasons mentioned by two of the cities are claimed changes in reporting software—with one citing FBI changes to the national system and the other citing conversion to new software at the local level.

As of the JRSA inventory, the state program (then the Mississippi Highway Patrol) was not NIBRS certified but three local agencies were: Biloxi, Byram, and Gulfport. Only Biloxi and Gulfport appear in the FBI's 2015 NIBRS tabulations as providing 12 months of NIBRS-format data.

Mississippi's Statistical Analysis Center is separately created in the School of Criminal Justice, University of Southern Mississippi, by executive order.

MISSOURI

- *Local law enforcement agencies required to submit crime statistics to state entity:* Missouri law (MO Rev Stat § 43.505, text originally enacted in 2001) designates the Department of Public Safety as the "central repository for the collection, maintenance, analysis and reporting of crime incident activity generated by law enforcement agencies in this state," and this repository/UCR system is directed to be "compatible with the national uniform crime reporting system operated by the Federal Bureau of Investigation." The department is obligated to "develop, operate and maintain an information system for the collection, storage, maintenance, analysis and retrieval of crime incident and arrest reports from Missouri law enforcement agencies." All law enforcement agencies in Missouri are obligated to "submit crime incident reports" and "any other crime incident information which may be required" to the department, subject to the penalty of becoming "ineligible to receive state or federal funds which would otherwise be paid to such agency for law enforcement, safety or criminal justice purposes." Though the legal wording is (almost) all phrased in terms of the Department of Public Safety, UCR functions are practically delegated to the Missouri Highway Patrol (within the department).
- *Transfer to FBI/national compilation required by law:* The state UCR program is mandated to "compile the statistical data and forward such data as required to the Federal Bureau of Investigation or the appropriate Department of Justice agency in accordance with the standards and procedures of the national system" (§ 43.505).
- *State maintains own incident-based reporting system:* The Missouri Incident Based Reporting System was NIBRS certified in 2005, and limited participation in the system has been achieved since that time. As of the JRSA

inventory, 22 agencies covering 11 percent of the state's population and 17 percent of its observed crime reported in NIBRS format; the FBI's 2015 NIBRS tabulation includes entries for 14 local Missouri agencies reporting 12 months of NIBRS data, the largest of those being the Kansas City Police Department.

- *Annual crime report required by law:* The Department of Public Safety is obligated to "annually publish a report on the nature and extent of crime and submit such report to the governor and the general assembly. Such report and other statistical reports shall be made available to state and local law enforcement agencies and the general public through an electronic or manual medium" (§ 43.505).
- *Other key features:* § 43.545 flags domestic violence as a priority reporting topic, and is the exception to other legislative language in that it directly addresses the Highway Patrol rather than the parent Department of Public Safety. Specifically, "the state highway patrol shall include all reported incidents of domestic violence as defined [elsewhere in state law], whether or not an arrest is made, in its system of reporting for compilation in the annual crime report. [All] incidents shall be reported on forms provided by the highway patrol and in a manner prescribed by the patrol."
- *State Statistical Analysis Center:* Same entity, Missouri State Highway Patrol.

MONTANA

The Montana Board of Crime Control serves both as the state's recognized UCR program and its Statistical Analysis Center. The board's authority regarding crime reporting is not spelled out in state law, but instead appears to derive solely from Montana Code Annotated § 44-4-301: "As designated by the governor as the state planning agency under the Omnibus Crime Control and Safe Streets Act of 1968, as amended, the board of crime control shall perform the functions assigned to it under that act. The board shall also provide to criminal justice agencies technical assistance and supportive services that are approved by the board or assigned by the governor or legislature."

Though "light" in terms of legal mandate, Montana was among the earliest states to establish a designated state UCR program and among the earliest adopters of NIBRS standards. The state's own incident-based reporting system, MTIBRS, was certified as fully NIBRS-compliant in 1992. According to the Montana Board of Crime Control (2016:2), MTIBRS differs from NIBRS in three ways. First, it includes 5 data elements not included in NIBRS specifications; these "pertain to domestic abuse, gang activity, gambling, mental health, and drug task force incidents." Second, MTIBRS "collects all combinations of offenses within an incident," so that MTIBRS data are "exported to NIBRS with a filter that excludes lesser included offenses" for some combinations, such as when an incident involves both an "aggravated assault" and a simple assault. NIBRS "only wants the aggravated assault but the state program will accept both offense codes." Finally, "MTIBRS collects administrative, offense, property, victim, offender and arrestee information on all offenses," not just NIBRS Group A offenses.

Montana Board of Crime Control (2016:11) notes that, "as of 2015, 100 Montana reporting law enforcement agencies out of 110 submitted 12 months of the MTIBRS data electronically. Four more agencies have submitted partial data for 2015. Six Montana agencies do not report data." The FBI's 2015 NIBRS tabulations show 12 months of contributed data from 94 Montana agencies; for NCS-X purposes, Montana is considered as complete NIBRS.

NEBRASKA

- *Local law enforcement agencies required to submit crime statistics to state entity:* The state Commission on Law Enforcement and Criminal Justice retains long-standing authority (NE Code § 81-1426) to "adopt and promulgate rules and regulations for the standardized collection, development, and maintenance of statistical information, records, and reports, including, but not limited to, the Uniform Crime Report." Further, "it shall be the duty of all public officers and agencies, the functions of which include the control, apprehension, trial, or correction of criminal offenders in the state, to provide such information, records, and reports, including, but not limited to, the Uniform Crime Report, as the commission determines relevant to its functions."
- *Transfer to FBI/national compilation not apparently referenced in law.*
- *State UCR program is NIBRS certified and permits reporting in either Summary or NIBRS format:* As of the JRSA inventory, 116 agencies covering 39 percent of Nebraska's population and 23 percent of its crime contributed NIBRS format; the FBI's 2015 NIBRS report shows 70 Nebraska agencies having reported 12 months of NIBRS data. The state program has been awarded an NCS-X Implementation grant, and the Omaha Police Department was among the local departments receiving NCS-X development grants in 2016.
- *Crime report required by law,* though not explicitly on annual cycle. § 81-1425 tasks the commission's executive director to "collect, develop, maintain, and analyze statistical information, records, and reports as the commission may determine relevant to its functions, including, but not limited to, the statistical information set forth in section 47-627" (described below).
- *Other key features:* Beginning in 2003 as part of a broad Community Corrections Act, Nebraska vested authority for creating a "uniform crime data analysis system" in the state's Department of Community Corrections

(§ 47-627)—creating a parallel data system focused on the state's correctional population. This system:

> shall include, but need not be limited to, the number of offenses, arrests, charges, probation admissions, probation violations, probation discharges, participants in specialized community corrections programs, admissions to and discharges from problem-solving courts, admissions to and discharges from the Department of Correctional Services, parole reviews, parole hearings, releases on parole, parole violations, and parole discharges. The data shall be categorized by statutory crime.

Another interesting feature of Nebraska's crime reporting law is the structure of the penalty for nonresponse for "willful or repeated failure by any public officers and agencies [to] submit the prescribed information, records, or reports, including the Uniform Crime Report:" such nonresponse "shall subject the agency or the administrator of the agency to a civil penalty of up to one hundred dollars per day for each day of violation," "recoverable by way of a civil suit brought against such agency or individually against the administrator."

- *State Statistical Analysis Center:* Same entity, Nebraska Crime Commission.

NEVADA

- *Local law enforcement agencies required to submit crime statistics to state entity:* The Central Repository for Nevada Records of Criminal History, under the Nevada Department of Public Safety, is directed by NV Rev Stat § 179A.078 to "establish [a] Uniform Program for Reporting Crimes that is designed to collect statistical data relating to crime or delinquency of children and to facilitate the collection and analysis of statistical data relating to crime at a central location." "Each agency of criminal justice and any other agency dealing with crime or delinquency of children" is obliged to collect and submit "reports and compilations of statistical data" required by the Department of Public Safety.
- *Transfer to FBI/national compilation required by law:* Per § 179A.075, the Central Repository is to tabulate the information from local agencies and "disseminate to federal agencies engaged in the collection of statistical data relating to crime information which is contained in the Central Repository."
- *State is in early planning for incident-based reporting system:* Listed in the JRSA inventory as having "no plans" for NIBRS implementation, Nevada has begun taking planning steps via NCS-X funding. The state program was awarded an NCS-X Planning grant, and the Las Vegas Metropolitan Police Department received NCS-X funding in 2016.
- *Annual crime report required by law:* § 179A.075 further obligates the UCR program to compile two annual reports "on or before July 1 of each year."

One, directed to the governor and state legislative leaders, is "a printed annual report containing the statistical data" on crime received during the preceding year; § 179A.175 requires that this report contain a section on bias-motivated crimes, those that "manifest evidence of prejudice based on race, color, religion, national origin, physical or mental disability, sexual orientation or gender identity or expression." The second report is a dedicated report "'containing statistical data about domestic violence in this State," addressed to specific legislative entities.

- *State Statistical Analysis Center:* Center for Analysis of Crime Statistics, Department of Criminal Justice, University of Nevada Las Vegas.

NEW HAMPSHIRE

- *Local law enforcement agencies required to submit crime statistics to state entity:* New Hampshire's UCR law was substantially revised in 2016 to etch into law that which had already been achieved: full NIBRS-format reporting. The post-July 2016 language of NH Rev Stat § 106-B:14-c is concise enough to state in its entirety (capitalization in original):

> I. Law enforcement agencies operating within New Hampshire shall submit crime reports to the department of safety, division of state police, based on the specifications prescribed by the Federal Bureau of Investigation's (FBI's) Criminal Justice Information Services' (CJIS) Uniform Crime Reporting (UCR) Program. The report shall be a compilation of the number of offenses and arrests on crimes occurring within their own jurisdictions, submitted in the manner and in the form specified by the FBI's National Incident Based Reporting System (NIBRS). Annual statistical reports shall be made available no later than July 1 of the following year, containing the number and nature of criminal offenses, arrests, and clearances, and any other data determined to be appropriate relating to the method, frequency, cause, and prevention of crime. The department of safety shall make such reports available to the governor, the general court, and all law enforcement agencies. Upon request for such reports to the division of state police, uniform crime report (UCR) unit, the department shall provide copies of the reports to other federal and state agencies and members of the public.
>
> II. A law enforcement agency having no records management system with NIBRS reporting capabilities shall be provided the means to report by the division of state police.

- *Transfer to FBI/national compilation required by law,* as implied by the text quoted above.
- *State has achieved full NIBRS compliance:* The state UCR program began accepting NIBRS-format data in 1997, and most local agencies received

NIBRS certification in the early 2000s. Norton and Delay (2012:7) indicate that "only 40 percent of communities participated in the NIBRS program" in New Hampshire in 1997; the state's largest and second largest cities, Manchester and Nashua, only began reporting NIBRS format in 2007 and 2005, respectively. By the time of the JRSA inventory, the state was deemed to have achieved NIBRS-format reporting for all agencies in the state, and it is deemed "complete NIBRS" for NCS-X purposes.

- *Annual crime report required by law,* as per the 2016 law quoted above. This would be a significant change from past practice: The independent New Hampshire Center for Public Policy Studies issued its own "Crime in New Hampshire" report in early 2012, based on 2011 data, observing that "New Hampshire has not published a comprehensive annual crime statistics report since the one released by the New Hampshire Department of Safety in 1993" (Norton and Delay, 2012:1).
- *State Statistical Analysis Center:* Separately vested in the Office of the Attorney General.

NEW JERSEY

- *Local law enforcement agencies required to submit crime statistics to state entity:* Establishment of a UCR program is explicitly a duty of the state attorney general, though the law permits these duties to be designated to the New Jersey State Police, within the Department of Law and Public Safety headed by the attorney general (NJ Rev State § 52:17B-5.1,5.2). This delegation to the New Jersey State Police is exercised in practice. Pursuant to § 17B-5.3, localities are required to participate in the state UCR program:

> All local and county police authorities shall submit a quarterly report to the Attorney General, on forms prescribed by the Attorney General, which report shall contain the number and nature of offenses committed within their respective jurisdictions, the disposition of such matters, . . . and such other information as the Attorney General may require, respecting information relating to the cause and prevention of crime, recidivism, the rehabilitation of criminals and the proper administration of criminal justice.

- *Transfer to FBI/national compilation not apparently referenced in law.*
- *State has no incident-based reporting system:* New Jersey is listed in the JRSA inventory as "testing" NIBRS compliance, rather than "no plans." The state program was awarded an NCS-X Planning grant to try to spur development.
- *Annual crime report required by law:* Per § 52:17B-5.4, the attorney general is tasked to "compile such statistics as he may deem necessary in order to present a proper classification and analysis of the volume and nature of crime

and the administration of criminal justice within this State;" § 52:17B-5.5 sets the frequency of this report as annual.

- *Other key features:* The enabling law highlights gang-related crime and bias-motivated crime as two key measurement priorities. § 52:17B-5.3 requires local departments to provide quarterly "information relating to criminal street gang activities within their respective jurisdictions," further stipulating that "a law enforcement officer who responds to an offense involving criminal street gang activity shall complete a gang related incident offense report on a form prescribed by the Superintendent of State Police" and submit that information "for inclusion in the Uniform Crime Report." In terms of bias-motivated crime, the quarterly reports from local agencies are to include "information relating to any offense directed against a person or group, or their property, by reason of their race, color, religion, gender, disability, sexual orientation, gender identity or expression, national origin, or ethnicity." An unusual characteristic of New Jersey's specification of annual reports is the directive that "classification and analysis" of collected crime data in the state must include "a section on the per capita crime statistics for a resort municipality"—that is, accounting for both the average-across-the-year population of such communities as well as the seasonal surges of population they experience (§ 17B-5.4).
- *State Statistical Analysis Center:* Also vested in the Department of Law and Public Safety.

NEW MEXICO

- *Local law enforcement agencies required to submit crime statistics to state entity:* As per NM Stat. § 29-3-11, the New Mexico Department of Public Safety "shall develop, operate and maintain a uniform crime reporting system and shall be the central repository for the collection, storage, retrieval and analysis of crime incident and arrest reports generated by all law enforcement agencies in this state. The system shall be operational as of January 1, 2008." All law enforcement agencies in the state are directed to "submit crime incident reports" and "any other crime incident information as may be required," and to "use uniform crime incident reporting as provided by the department for all incidents and arrests," though no penalty is specified.
- *Transfer to FBI/national compilation required by law:* The enabling statute obliges the department to "compile statistical data and forward such data as required to the federal bureau of investigation or the appropriate department of justice agency in accordance with standards and procedures of the national system" (NM Stat. § 29-3-11; capitalization original)
- *State has no incident-based reporting system,* but was listed in the JRSA inventory as "testing" NIBRS possibilities. Though the state UCR program

has not received NCS-X funds, the Albuquerque Police Department did receive such funding in 2016.

- *Annual crime report required by law:* An annual report "on the nature and extent of crime in New Mexico" is required to be completed "in conjunction with the New Mexico sentencing commission," and submitted to the governor and legislature annually (NM Stat. § 29-3-11).
- *State Statistical Analysis Center:* Separately vested in Institute for Social Research, University of New Mexico.

NEW YORK

- *Local law enforcement agencies not directly required to submit crime statistics to state entity:* New York Executive Law § 837–4(c) directs the State Division of Criminal Justice Services to "collect and analyze statistical and other information and data with respect to the number of crimes reported or known to police officers or peace officers, the number of persons arrested for the commission of offense, the offense for which the person was arrested, the county within which the arrest was made and the accusatory instrument filed, [and] the disposition of the [accusatory instrument within the justice system]." However, that clause is nested in language that stops short of mandating such data collection; rather, it is supposed to be done "in cooperation with the state administrator of the unified court system as well as any other public or private agency" (Executive Law § 837–4).
- *Transfer to FBI/national compilation encouraged by law:* The Division is directed to "supply data, upon request, to federal bureaus or departments engaged in collecting national criminal statistics" (Executive Law § 837–4(d)).
- *State maintains own incident-based reporting system:* The New York State Incident-Based Reporting (NYSIBR) system was said, at the time of the JRSA inventory, to cover 22 percent of the state's population and 17 percent of its crime. NYSIBR and NIBRS content do not directly align; historically, NYSIBR did not include 4 of the variables required by NIBRS, but added nearly 15 data elements not coded into NIBRS (including the level of injury and medical treatment received by the victim). Work is taking place on reconciling the differences between the systems, hence its designation in the JRSA inventory as "testing." The New York State Police was awarded NCS-X development funds, and the state program received an NCS-X Implementation grant.
- *Quarterly crime report required by law:* The division is required to present "a quarterly report containing the statistics and other information" to the governor and state legislative leaders on a quarterly basis (Executive Law § 837-a-2).

- *Other key features:* A final clause of the basic crime reporting statute stipulates that, "where a [firearm] comes into the custody of police officers or peace officers in the course of an investigation of such crime or offense," additional detailed information on the firearm be collected—including "the make, model type, caliber and magazine or cylinder capacity of any such firearm and whether possession of such firearm by the defendant is licensed or unlicensed and if confiscated at arrest, the style and manufacturer of any ammunition" (New York Executive Law § 837–4(c)). New York statute is also unique in flagging that information on some particular offense classes be collected by the Division of Criminal Justice Services, subject to the same "in cooperation with" the courts and other agencies clause: specifically, "environmental crimes and offenses" including violations of state environmental conservation laws, "crimes and offenses committed against employees of the city of New York responsible for enforcing certain regulations in such city," hate crimes, and insurance fraud (Executive Law § 837–4-a, 4-b, and 4-c, and § 837-a–1, respectively).
- *State Statistical Analysis Center:* Same entity, Division of Criminal Justice Services.

NORTH CAROLINA

- *Local law enforcement agencies required to submit crime statistics to state entity:* Uniform crime reporting in North Carolina derives from a single, expressive clause in law. Since 1985, NC General Statutes § 143B-906 has held that:

  > It shall be the duty of the State Bureau of Investigation to receive and collect criminal information, to assist in locating, identifying, and keeping records of criminals in this State, and from other states, and to compare, classify, compile, publish, make available and disseminate any and all such information to the sheriffs, constables, police authorities, courts or any other officials of the State requiring such criminal identification, crime statistics and other information respecting crimes local and national, and to conduct surveys and studies for the purpose of determining so far as is possible the source of any criminal conspiracy, crime wave, movement or cooperative action on the part of the criminals, reporting such conditions, and to cooperate with all officials in detecting and preventing.

  The law makes no explicit mention of a requirement to submit data to the state program, and the program is described as purely voluntary.
- *Transfer to FBI/national compilation not directly specified by law.*
- *State maintains own incident-based reporting system:* North Carolina has maintained an incident-based reporting system for decades—since 1980, according to descriptive material on the agency's website that appears to have been adapted from an annual report circa 1999 (the site, accessible as of July 2017, is http://crimereporting.ncsbi.gov/Introduction.aspx). That website

material suggests that 430 of 460 reporting agencies in the state used incident-based format in 1999. Both the North Carolina state UCR program and the Charlotte Police Department have received NCS-X funding.

- *Annual crime report required by law:* The law actually mandates a biennial report, even though it has been issued annually. Specifically, the bureau is required "to make scientific study, analysis and comparison from the information so collected and correlated with similar information gathered by federal agencies, and to provide the Governor and the General Assembly with the information so collected biennially, or more often if required by the Governor" (NC General Statutes § 143B-902).
- *State Statistical Analysis Center:* None, at present.

NORTH DAKOTA

- *Local law enforcement agencies required to submit crime statistics to state entity:* Under North Dakota law, "criminal justice training and the collection of statistics must be conducted by the office of attorney general." "To assist in controlling crime in the state through the use of reliable statistics relating to crimes and criminal activity, the Attorney General may obtain [from criminal justice agencies] all information the Attorney General deems necessary;" . Moreover, the law makes it "the duty of the officials to furnish the information ... within thirty days of the request on whatever forms or in whatever manner the Attorney General may prescribe" (ND Century Code § 12-62-01.1). The operation of the UCR program in North Dakota is delegated to the Bureau of Criminal Investigation.
- *Transfer to FBI/national compilation not mentioned in law.*
- *State maintains own incident-based reporting system:* North Dakota law enforcement agencies have primarily reported incident-based format data to the state since at least 1999, with a small number of agencies reporting SRS format through the early 2000s. The state's annual crime report for 2011, covering data collected in 2010, was the first with full incident-based participation.[2]
- *Annual crime report not apparently specified by law* (but still completed annually).
- *State Statistical Analysis Center:* Same entity, Bureau of Criminal Investigation.

OHIO

[2]The archive of annual reports is accessible at https://attorneygeneral.nd.gov/public-safety/crime-data/crime-reports.

- *Local law enforcement agencies encouraged to submit crime statistics to state entity:* Ohio made use of a lapse in its intrastate UCR program to etch incident-based crime reporting into its law. For years, some local jurisdictions continued to report SRS-format data while others switched to the Ohio Incident Based Reporting System (OIBRS). Specifically, Ohio's original UCR Program "remained in place [from 1976 to] 1981 when its cancellation was dictated by state budget. Since that time local law enforcement agencies have returned to the old practice of submitting UCR information directly to the FBI" (Ohio Department of Public Safety, 2017:5). Cognizant of the developments that would produce NIBRS, the state set about reestablishing its UCR program in 1986; aided by a BJS grant, OIBRS began active development work in 1989. Hence, when the program was reauthorized, it was established as an incident-based–only system: The Division of Criminal Justice Services, in the Department of Public Safety, is tasked to "develop and maintain the Ohio incident-based reporting system to facilitate the sharing of information with the federal bureau of investigation and participating law enforcement agencies in Ohio" (Ohio Rev Code § 5502.62). Under the law, local contribution of data to OIBRS is, strictly speaking, voluntary; any law enforcement agency that receives either "federal criminal justice grants or state criminal justice information system general revenue funds through the division" is required by law "to participate in OIBRS" *or* to participate "in the uniform crime reporting program of the federal bureau of investigation" (lower case in original).
- *Transfer to FBI/national compilation required by law:* As conveyed in the text above, Ohio's crime-reporting law is unusual in that participation in the national UCR program is reinforced more strongly than in the state-level program.
- *State maintains own incident-based reporting system:* OIBRS, as denoted in law. For NCS-X purposes, Ohio is considered to be ineligible for grant funding as it is more than 80 percent NIBRS compliant; the JRSA inventory found that 83 percent of the state's population and 71 percent of observed crime were covered by NIBRS-format reporters.
- *Annual crime report not required by law:* There is no apparent mandate for an annual report on the state of crime, though the division is obligated by its authorizing law to "provide assistance, advice, and reports requested by the governor, the general assembly, or the federal bureau of investigation" (Ohio Rev Code § 5502.62).
- *State Statistical Analysis Center:* The same law that reestablished Ohio's UCR program granted the Division of Criminal Justice Services with principal authority to "collect, analyze, and correlate information and data concerning the criminal justice system in the state;" it is both the state SAC and UCR program (Ohio Rev Code § 5502.62).

OKLAHOMA

- *Local law enforcement agencies required to submit crime statistics to state entity:* State law authorizes the Oklahoma State Bureau of Investigation (SBI) to establish a UCR program, and "all state, county, city and town law enforcement agencies shall submit reports to the [SBI] on forms prescribed by the Bureau." Specifically, "the reports shall contain the number and nature of offenses committed within their respective jurisdictions, the disposition of such matters, and such other information as the Bureau may require, respecting information relating to the cause and prevention of crime, recidivism, the rehabilitation of criminals and the proper administration of criminal justice" (74 OK Stat. § 74-150.10C).
- *Transfer to FBI/national compilation not apparently referenced by law.*
- *State maintains own incident-based reporting system,* apparently referred to simply as the State Incident-Based Reporting system. The state of Oklahoma is NIBRS certified, and the JRSA inventory suggested that 28 percent of its population and 11 percent of crime was covered by NIBRS-format reporters. The state program was awarded an NCS-X Implementation grant.
- *Crime report required by law:* The SBI is directed to "compile such statistics as the Director may deem necessary in order to present a proper classification and analysis of the volume and nature of crime and the administration of criminal justice within this state" in a report, but there is no stated frequency (74 OK Stat. § 74-150.10D).
- *Other key features:* Short of withholding grant eligibility, Oklahoma's penalty for "refusal or persistent failure of any law enforcement agency to submit reports required by this section" is "discontinued access to Bureau information and assistance" (74 OK Stat. § 74-150.10E).
- *State Statistical Analysis Center:* "The Oklahoma State Bureau of Investigation shall be the entity recognized by the Bureau of Justice Statistics as the Statistical Analysis Center" (22 OK Stat. § 22-1517).

OREGON

- *Local law enforcement agencies required to submit crime statistics to state entity:* Oregon's crime-reporting legislation explicitly defines the national UCR content as a *subset* of information required from local law enforcement agencies. As per OR Rev. Stat. § 181A.225(1):

> All law enforcement agencies shall report to the Department of State Police statistics concerning crimes:
> (a) As directed by the department, for purposes of the Uniform Crime Reporting System of the Federal Bureau of Investigation.
> (b) As otherwise directed by the Governor concerning general criminal categories of criminal activities but not individual criminal records.

(c) Motivated by prejudice based on the perceived race, color, religion, national origin, sexual orientation, marital status, political affiliation or beliefs, membership or activity in or on behalf of a labor organization or against a labor organization, physical or mental disability, age, economic or social status or citizenship of the victim.

(d) And other incidents arising out of domestic disturbances.

- *Transfer to FBI/national compilation not directly specified by law*—but is certainly consistent with the language cited above.
- *State maintains own incident-based reporting system, and discontinued Summary-format reporting in January 2017:* The state crime reporting program accepts data submissions in either straight NIBRS format or in ONIBRS format—the Oregon National Incident-Based Reporting System. The principal difference in content between the two is that ONIBRS covers certain regulatory offenses not handled in NIBRS: some traffic crimes, fish and game regulatory violations, and marine violations (Oregon Uniform Crime Reporting Program, 2017). As of the JRSA inventory, 29 percent of Oregon's population and 28 percent of its crime were covered by incident-based reporting agencies, and the state program is itself NIBRS certified. A prominent exception to incident-based reporting—the Portland Police Department—was awarded NCS-X development funds, while the state UCR program is deemed a "state-led" conversion for NCS-X purposes.
- *Annual crime report required by law:* The enabling law requires general statistical crime reports on both a quarterly and annual basis, reports on statistics of domestic disturbances on both a quarterly and annual basis, and an "annual public report" on bias-motivated crimes. In addition, the Department of State Police is directed to complete "special reports" as requested by the governor (OR Rev. Stat. § 181A.225(2)).
- *State Statistical Analysis Center:* Separately vested in Oregon Criminal Justice Commission.

PENNSYLVANIA

- *Local law enforcement agencies required to submit crime statistics to state entity:* A Pennsylvania Uniform Crime Reporting program was first established in 1970, placed under authority of the attorney general. In 1973, authority was transferred to the Pennsylvania State Police, and current authority stems from the Uniform Crime Reporting Act enacted in 2004. Under that law, the state UCR program "shall be established under the direction, control and supervision of the Pennsylvania State Police," which "shall have the power and duty to promulgate such rules and regulations" as necessary "to collect and gather the information from State, county and local law enforcement agencies" (18 P.S. § 20.503, 2014). Prior to the 2004 act, contribution of data to the state program was voluntary, but a lengthy section of the 2004 legislation established penalties for failure to comply with monthly offense reporting—principally, removal of eligibility for distribution of funds from the collection of fines (§ 20.504)

- *Transfer to FBI/national compilation is apparently not mentioned in law.*
- *State does not maintain incident-based reporting system:* As of the JRSA inventory, the state was said to be "testing" NIBRS compliance issues. The FBI's 2015 NIBRS tabulations suggest 12 months of NIBRS-format submissions by only 22 (small) Pennsylvania agencies. The state program was awarded NCS-X Implementation monies, and the Philadelphia Police Department was awarded NCS-X development grant funds in 2016.
- *Annual crime report required by law:* Drawing from 2004 language, Pennsylvania's requirement of an annual report in that it specifies that it be in electronic format—"the Pennsylvania State Police shall electronically publish an annual report containing results of the information gathered and collated for access by the Governor, the General Assembly, the Attorney General and the public" (§ 20.506).
- *State Statistical Analysis Center:* Separately vested in Office of Research, Evaluation, and Strategic Policy Development, Pennsylvania Commission on Crime and Delinquency.

RHODE ISLAND

- *Local law enforcement agencies required to submit crime statistics to state entity:* "A uniform crime reporting system shall be established under the direction, control, and supervision of the superintendent of state police. The superintendent of state police shall have the power and duty, by any rules and regulations that he or she may deem necessary to collect and gather any information from local police departments and the enforcement division of the department of environmental management that may be and is prescribed in this chapter" (RI Gen Laws § 12-24-1). § 12-24-2 expands on this, holding that "the police department of each city and town and the enforcement division of the department of environmental management, once each quarter, upon a date and form prescribed and furnished by the superintendent of state police, shall forward to the division of state police a crime report."
- *Transfer to FBI/national compilation required by law:* The statewide compilation of agency-submitted data must be "available to any governmental law enforcement agency in the state, the judiciary committees of the senate and the house of representatives, and the Federal Bureau of Investigation, upon request" (§ 12-24-2).
- *State maintains own incident-based reporting system:* The Rhode Island Incident Based Reporting System (RIBR) system evidently began operations in 2005, winning NIBRS certification in 2006. Though the JRSA profile lists

Rhode Island as fully NIBRS-compliant with 100 percent of jurisdictions reporting in the format, it is deemed as just "> 80% NIBRS" for NCS-X purposes.

- *Annual crime report not apparently referenced in law.*
- *State Statistical Analysis Center:* Rhode Island Department of Public Safety, Grants Administration Office.

SOUTH CAROLINA

- *Local law enforcement agencies required to submit crime statistics to state entity:* The South Carolina Code of Laws grants general information-gathering authority to the South Carolina State Law Enforcement Division (SLED); the state's Code of Regulations builds on that authority to define roles more completely. Specifically, SC Code of Regulations § 73-21(4) creates a UCR Department that "has the responsibility for processing, analyzing, coding and compiling incident, supplemental, and booking reports received from law enforcement agencies, whether such reports are submitted on paper or by automated means." § 73-30 mandates that "every law enforcement agency must send SLED a copy of each report made by any officer during the performance of his duties in responding to reported criminal violations, ... regardless of the degree of seriousness of the reported criminal activity." These incident reports "must be recorded on standard forms approved by SLED" and "must be forwarded to SLED on a regular and timely basis, but not less often than once weekly"; incident reports "during any month must arrive at SLED no later than the fifth day of the following month."
- *No apparent explicit legal mandate for data transfer to the FBI/national compilation:* However, the state regulations are replete with references to FBI UCR standards, and transfer to the national program is certainly consistent with the spirit of the regulation text.
- *State maintains own incident-based reporting system:* The South Carolina Incident-Based Reporting System (SCIBRS) was the first state incident-based reporting system, serving as a model/demonstration for NIBRS formatting and beginning operations in 1991. Beginning with its 2013 annual report, South Carolina began to feature exclusively incident-based reporting data (rather than SCIBRS-format data converted to Summary Reporting System-style counts). Accordingly, it is listed in the JRSA inventory as fully NIBRS compliant.
- *Annual crime report not apparently referenced in law.*
- *Other key features:* In defining its role in the crime reporting process, SLED assumes a strong role in ensuring data quality and implementing consistent editing/classification routines. Per SC Code of Regulations § 73-21(4), the UCR department:

> will classify and count incident, supplemental, and booking reports submitted by other agencies according to procedures defined by the International Association of Chiefs of Police Committee on Uniform Crime Reports, the Uniform Crime Records Committee of the National Sheriffs Association, the Uniform Crime Reports Section of the Federal Bureau of Investigation and the State Law Enforcement Division. The Uniform Crime Reporting Department will assure through training and quality control measures that all automated incident, supplemental, and arrest data submitted to the State Uniform Crime Reporting program are classified and counted according to these procedures.

(This responsibility is reaffirmed in § 73-30.)

- *State Statistical Analysis Center:* Office of Justice Programs, Department of Public Safety.

SOUTH DAKOTA

- *Local law enforcement agencies required to submit crime statistics to state entity:* The attorney general of South Dakota, who is also designated by law as director of the state Bureau of Criminal Statistics, is directed to (SD Codified Laws § 23-6-4):

> collect and compile information, statistical and otherwise, which will, as far as practicable, present an accurate survey of the number and character of crimes committed in the state, the extent and character of delinquency, [and the administration of justice]. He shall include such information as may be useful in the study of crime and delinquency and the causes thereof, for the administration of criminal justice, and for the apprehension, punishment and treatment of criminal offenders.

Pursuant to § 23-6-11, "every person having custody or charge of public or official records or documents, from which information is sought for the purposes of this chapter, shall grant to the director, or to any person deputized by him, access thereto."[3] The state's designated UCR program is housed in the Division of Criminal Investigation.

- *Transfer to FBI/national compilation is not directly referenced in law*—but such is certainly consistent with the law, which includes a quirk. The legal language establishing state crime statistics dates back to South Dakota's Uniform Criminal Statistics Act enacted in 1939, at which time there remained some uncertainty as to where authority for national crime statistics would reside. Hence, South Dakota statute is perhaps unique in referencing the U.S. Census Bureau's historical role in crime statistics: § 23-6-6 requires the Bureau of Criminal Statistics to "promulgate classifications and shall

[3]That said, the South Dakota attorney general's website states bluntly that "crime reporting is voluntary" http://atg.sd.gov/OurOffice/Departments/DCI/SAC/NIBRS.aspx.

prepare forms for the statistical classification of crimes, of offenders, of their punishment and treatment and of all other pertinent information, to conform, as far as practicable, with those promulgated by the appropriate agency in the United States Department of Justice, and by the Federal Bureau of the Census."

- *State is fully NIBRS compliant:* As described by the attorney general's website (http://atg.sd.gov/OurOffice/Departments/DCI/SAC/NIBRS.aspx), the South Dakota UCR program was NIBRS certified in 2001. In 2006, a new web-based NIBRS tool was made available to South Dakota law enforcement agencies to encourage adoption of the new format; the move enabled the state to become NIBRS-only two years later. The JRSA profile and NCS-X planning tabulation both identify South Dakota as 100 percent NIBRS, yet the attorney general's website says "approximately 95% of the state's population is covered by NIBRS," without specifying what constitutes the other 5 percent.
- *Annual crime report required by law:* Reports on "the crime situation in this state" are required "annually, and at such other times as he may determine," from the attorney general. These reports are intended to "point out what he considers to be significant features regarding crime, the administration of criminal justice and the punishment or treatment of criminals, and may recommend such measures as he may consider desirable or constructive with reference thereto" (§ 23-6-10).
- *Other key features:* Most clauses in the section of code regarding the collection of crime statistics—the basic responsibility of the attorney general for collection, and even the promulgation of classifications of crime—include language that make violations of the clauses crimes in themselves: "a violation of this section is a Class 2 misdemeanor" (§ 23-6-4 and elsewhere).
- *State Statistical Analysis Center:* Same entity, Division of Criminal Investigation.

TENNESSEE

- *Local law enforcement agencies required to submit crime statistics to state entity:* Tennessee Code § 38-10-101 directs the Tennessee Bureau of Investigation (TBI) to "establish a system of intrastate communication of vital statistics and information relating to crime, criminals, and criminal activity." "All state, county, and municipal law enforcement and correctional agencies, and courts, shall submit [to the TBI] reports setting forth their activities in connection with law enforcement and criminal justice, including uniform crime reports,"[4] as required by § 38-10-102. Moreover, § 38-10-103 grants the

[4]New law effective January 1, 2017, continues this phrase to include "reports of law-enforcement-related deaths."

TBI authority to "adopt and promulgate rules and regulations prescribing the form, general content, time, and manner of submission of the uniform crime reports," holding that such rules "shall have the force and effect of law."

- *Transfer to FBI/national compilation required by law:* The law does not explicitly reference information sharing with the FBI, but the rules promulgated under § 38-10-103 state "compiling and transmitting the data to the FBI monthly for all reporting agencies" as a core responsibility of the TBI (Rules of the TBI § 1395-1-2-.03), and note legislative intent for sharing of the data "by agreements between the Tennessee Bureau of Investigation, Federal Bureau of Investigation and National Law Enforcement Telecommunications Systems" (Rules of the TBI § 1395-1-1.02).
- *State maintains own incident-based reporting system:* Tennessee was a relatively early adopter of its own incident-based reporting system, TIBRS; the original rules defining TIBRS administration were apparently filed with the Tennessee Secretary of State in 2001 and took effect in 2002.
- *Annual crime report required by law:* By Tennessee Code § 38-10-104, the TBI "shall correlate the reports submitted" by local agencies "and shall compile and submit to the governor and the general assembly an annual report based on such reports." Said report "shall contain demographic data, including the percentage of suspects, victims, and convicted offenders based on race, gender, age, nationality, and any other appropriate demographic, as determined by the director of the Tennessee bureau of investigation."
- *Other key features:* As with some other states, failure to report crime data pursuant to the laws and regulations is subject to penalty in Tennessee: "Any officer or official mentioned in this chapter who shall have been notified and refuses to make any report or do any act required by any provision of this chapter shall be deemed guilty of nonfeasance of office and subject to removal therefrom" (§ 38-10-105).
- *State Statistical Analysis Center:* Same entity, Tennessee Bureau of Investigation.

TEXAS

- *Local law enforcement agencies required to submit crime statistics to state entity:* Texas Government Code § 411.042(b)(2) vests the Texas Department of Public Safety's Bureau of Identification and Records with authority to:

  > collect information concerning the number and nature of offenses reported or known to have been committed in the state and the legal steps taken in connection with the offenses, and other information useful in the study of crime and the administration of justice, including information that enables the bureau to create a statistical breakdown of: (A) offenses in which family and violence was involved; (b) [sexual assault and aggravated sexual assault offenses defined under Texas Penal Code]; and (C) [trafficking of persons and 'compelled prostitution' offenses under Texas Penal Code].

  The clause implying mandatory response to the state reads: "A law enforcement agency shall report offenses under Section 22.011 or 22.021, Penal Code, to the department in the form and manner and at regular intervals as prescribed by rules adopted by the department" (Texas Government Code § 411.042(i)). An apparent quirk in the law is that the referenced Penal Code sections cover only sexual assault offenses.
- *Transfer to FBI/national compilation encouraged by law:* The Bureau is directed to "cooperate with identification and crime records bureaus in other states and the United States Department of Justice" (Texas Government Code § 411.042(b)(4)).
- *Legally established goal of full incident-based reporting:* In September 2015, Texas enacted legislation "establish[ing] a goal that, not later than September 1, 2019, all local law enforcement agencies: (1) will have implemented an incident-based reporting system that meets the reporting requirements of the National Incident-Based Reporting System of the Uniform Crime Reporting Program of the Federal Bureau of Investigation; and (2) will use [that system] to submit to the department information and statistics concerning criminal offenses committed in the jurisdiction of the local law enforcement agency." The act further requires that the Department of Public Safety report to the legislature by January 2017 on progress in meeting this goal (Texas Government Code § 411.054). The first such progress report indicated that 97 Texas law enforcement agencies (covering 14.4 percent of the state's population) were NIBRS-compliant, with an additional 23 agencies about to come online—this relative to the total of 1,068 agencies reporting data to the state's UCR program (in Summary Reporting System or NIBRS format). For NCS-X purposes, the state program is considered a "state-led" initiative; however, the Dallas and San Antonio Police Departments were awarded NCS-X development funds in 2016.
- *Annual crime report not apparently referenced in law.*

- *Other key features:* For sexual assault offenses, Texas law (prior to the incident-based reporting legislation described above) prescribed the collection of additional incident-level detail. Specifically, Texas Government Code § 411.042(h) mandates that "information collected to perform a statistical breakdown of [these] offenses [must] include information indicating the specific offense committed and information regarding: (1) the victim; (2) the offender and the offender's relationship to the victim; (3) any weapons used or exhibited in the commission of the offense; and (4) any injuries sustained by the victim."
- *State Statistical Analysis Center:* None, at present.

UTAH

- *Local law enforcement agencies required to submit crime statistics to state entity:* UT Code § 53-10-202(2) directs the Bureau of Criminal Identification within the Utah Department of Public Safety to "establish a statewide uniform crime reporting system that shall include: (a) statistics concerning general categories of criminal activities; (b) statistics concerning crimes that exhibit evidence of prejudice based on race, religion, ancestry, national origin, ethnicity, or other categories that the division finds appropriate; and (c) other statistics as required by the Federal Bureau of Investigation." Section 53-10-206 holds that the bureau "may require all peace officers, the warden of the state prison, the keeper of any jail or correctional institution, or superintendent of the state hospital to obtain information that will aid in establishing the records required to be kept."
- *Transfer to FBI/national compilation is not explicitly required by law:* The passage just quoted, hewing the content of state crime reporting to data items "required" by the FBI, would seem to imply "mandatory" turnover of the data.
- *State maintains own incident-based reporting system:* The Utah Incident Based Reporting System began work in 1991, with 75 initial contributors; the JRSA inventory suggests incident-based reporting coverage of 81 percent of the state's population and 86 percent of crime. Despite being listed above the 80 percent mark in the JRSA profile, the Utah program was eligible for NCS-X funds and received an Implementation grant.
- *Annual crime report required by law:* The Bureau of Criminal Identification is required to "publish an annual report concerning the extent, fluctuation, distribution, and nature of crime in Utah" (UT Code § 53-10-202(5)).
- *Other key features:* Utah is one of very few states whose legislation includes an explicit disclaimer on the use of crime data: "The data acquired under the statewide uniform crime reporting system shall be used only for research or statistical purposes and may not contain any information that may reveal the

identity of an individual victim of a crime" (UT Code § 53-10-205). Likewise, Utah's code makes it incumbent on local communities and authorities to provide the means for their law enforcement agencies to be able to supply data: "All governing boards or commissions of each city, town, county, or correctional institution of the state shall furnish the appropriate officials with supplies and equipment necessary to perform the duties prescribed in this part" (UT Code § 53-10-212).

- *State Statistical Analysis Center:* Utah Commission on Criminal and Juvenile Justice.

VERMONT

- *Local law enforcement agencies required to submit crime statistics to state entity:* The Vermont Crime Information Center in the Department of Public Safety is designated as the official repository of criminal justice information, but the basic enabling legislation casts the relationship between the state and local agencies as a cooperative. "All state departments and agencies, municipal police departments, sheriffs and other law enforcement officers shall cooperate with and assist the center in the establishment of a complete and uniform system of records relating to the commission of crimes, arrests," and so forth (20 V.S.A. § 2053), and the center is empowered to suggest "uniform forms for the reporting of the commission of crimes" and other data items (20 V.S.A. § 2054). Local agency failure to contribute data is subject to penalty; see below.
- *Transfer to FBI/national compilation encouraged by law:* Similar "cooperate" language characterizes data transfer and sharing: "The center shall cooperate with other state [and local agencies] and with federal and international law enforcement agencies to develop and carry on a uniform and complete state, interstate, national and international system of records of criminal activities and information" (20 V.S.A. § 2053)
- *State is fully NIBRS compliant:* All of Vermont's law enforcement agencies appear to have won NIBRS certification by mid-2004, per the JRSA inventory
- *Annual crime report required by law:* "From time to time but at least annually, the center shall publish information relating to criminal activity, arrests, convictions and such other information as the commissioner deems relevant" (20 V.S.A. § 2057).
- *Other key features:* "A department, agency or law enforcement officer who fails to comply with the regulations adopted by the director [for crime reporting] with respect to the use, completion or filing of the uniform forms, after notice of failure to comply, shall be fined not more than $100.00. Each such failure shall constitute a separate offense" (20 V.S.A. § 2054).

- *State Statistical Analysis Center:* Services provided by contract to the Department of Public Safety by Crime Research Group, Inc.

VIRGINIA

- *Local law enforcement agencies required to submit crime statistics to state entity:* The superintendent of the Virginia State Police "shall establish, organize, equip, staff and maintain within the Department of State Police, at such departmental locations as the Superintendent may direct, [a] uniform crime reporting system for the purpose of receiving, compiling, classifying, analyzing and publishing crime statistics of offenses known, persons arrested, and persons charged and other information pertaining to the investigation of crime and the apprehension of criminals, as hereinafter provided" (Va. Code Ann. § 52-29, 2015). Section 52-28 makes it the duty of "all Commonwealth, county and municipal law-enforcement agencies" to provide "all periodic uniform crime reports setting forth their activities in connection with law enforcement." Further, § 52-26 authorizes the State Police to "maintain liaison and to cooperate with law-enforcement and criminal justice agencies of all counties, cities and towns and all other agencies, departments, and institutions of the Commonwealth, other states and of the United States in order to develop and carry on a comprehensive uniform crime reporting program for the Commonwealth."
- *Transfer to FBI/national compilation required by law:* The State Police's data collection authority "includ[es], but [is] not limited to, part I and part II offenses as set out by the Federal Bureau of Investigation" (§ 52-29); moreover, the State Police "shall correlate reports submitted to it and shall compile and submit reports to the Federal Bureau of Investigation on behalf of all agencies of the Commonwealth, as may be required by the federal standards for the uniform crime reporting program" (§ 52-30).
- *State maintains own incident-based reporting system:* A very early adopter of incident-based reporting, the Virginia State Police stopped accepting Summary-format, hardcopy reports in 2000, which was the first year in which the annual *Crime in Virginia* report reflected all incident-based reporting.
- *Annual crime report required by law:* The enabling law for the state UCR program references the "annual Crime in Virginia report," requiring that information on gang-related crimes and officer-involved shootings, in particular, be included in it (§ 52-28.1 and 52-28.2).
- *State Statistical Analysis Center:* Separately vested in Criminal Justice Research Center, Department of Criminal Justice Services.

WASHINGTON

- The Uniform Crime Reporting Program in Washington is described as a "cooperative" between the Washington Association of Sheriffs and Police Chiefs (WASPC) and the state's Criminal Justice Training Commission (Washington Association of Sheriffs and Police Chiefs, 2017). WASPC is itself a legally recognized "combination of units of local government," a consortium of local law enforcement agencies operating in the state (WA Rev Code § 36.28A.010). WASPC is the designated state UCR Program, and one passage of law characterizes the Criminal Justice Training Commission as maintaining "its contract for collection of statewide crime data" with the WASPC (WA Rev Code § 10.99.130(12)).
- *Local law enforcement agencies not required to submit crime statistics to state entity:* The transfer of data to WASPC is characterized as being entirely voluntary.
- *Transfer to FBI/national compilation not apparently mentioned in law*, though the Washington Association of Sheriffs and Police Chiefs (2017) crime report for 2016 does indicate that it is part of the unit's mission to forward crime data to the FBI.
- *State is largely incident-based reporting compliant:* According to the (Washington Association of Sheriffs and Police Chiefs, 2017:2–3):

    > The Washington State UCR Program was activated in October 1979 and collection of the monthly SRS data from contributing agencies began in January 1980. In December 2006, Washington State was certified by the FBI to officially collect and submit NIBRS data. In January 2008, based on a 2007 advisory committee recommendation, the WASPC Executive Board made the decision to set the goal date for Washington State law enfo rcement agencies to convert from the SRS method to NIBRS by January 1, 2012. … The majority of Washington State agencies submit via the NIBRS method and our goal is 100% NIBRS participation in the near future.

    The JRSA inventory suggests that incident-based reporting covered 45 percent of the state's population and 34 percent of crime; the state program was awarded an NCS-X Implementation grant.
- *Other key features:* At least two major crime classes are called out for special reporting by state law. A subsection of the passage of law authorizing WASPC directs the association to serve as repository for information on "malicious harassment"—acts of assault or threat that are committed based on offender's "perception of the victim's race, color, religion, ancestry, national origin, gender, sexual orientation, or mental, physical, or sensory handicap" (WA Rev Stat § 36.28A.030, 9A.36.080). Further, effective in 1994, WASPC was directed to collect data for major offense types to differentiate domestic violence incidents from others, including information on "violations of

the provisions of a protection order or no-contact order" (WA Rev Stat § 10.99.030).

- *State Statistical Analysis Center:* By state executive order, housed in Washington Office of Financial Management.

WEST VIRGINIA

- *Local law enforcement agencies required to submit crime statistics to state entity:* As per WV Code § 15-2-24, "all state, county and municipal law-enforcement agencies shall submit to" the West Virginia State Police "uniform crime reports setting forth their activities in connection with law enforcement. It shall be the duty of the bureau to adopt and promulgate rules and regulations prescribing the form, general content, time and manner of submission of such uniform crime reports."
- *Transfer to FBI/national compilation not apparently mentioned in law.*
- *State maintains own incident-based reporting system:* State is fully incident-based report compliant, and began accepting only West Virginia Incident-Based Reporting System-format data from local departments in 1999
- *Semiannual crime report required by law:* "The bureau shall correlate the reports submitted to it and shall compile and submit to the Governor and the Legislature semiannual reports based on such reports. A copy of such reports shall be furnished to all prosecuting attorneys and law- enforcement agencies" (WV Code § 15-2-24).
- *State maintains own incident-based reporting system and is fully NIBRS compliant:* As per a descriptive page (http://www.djcs.wv.gov/ORSP/SAC/Pages/WVIBRS.aspx) and the JRSA inventory, the West Virginia Incident-Based Reporting System was an early adherent to NIBRS standards, so much so that the state program "began only accepting data in the WVIBRS format as of January 1, 1999."
- *Other key features:* Violation of crime reporting routines is itself technically considered a crime. "Willful or repeated failure by any state, county or municipal law-enforcement official to submit the uniform crime reports required by this article shall constitute neglect of duty in public office" (WV Code § 15-2-24).
- *State Statistical Analysis Center:* Housed in the West Virginia Office of Research and Strategic Planning, a unit of the Division of Justice and Community Services within the state Department of Military Affairs and Public Safety.

WISCONSIN

- *Local law enforcement agencies required to submit crime statistics to state entity:* The Wisconsin Department of Justice is directed to "collect information concerning the number and nature of offenses known to have been committed in this state and such other information as may be useful in the study of crime and the administration of justice. The department of justice may determine any other information to be obtained regarding crime and justice system statistics" (WI Stat. § 165.845(1)(a), 2016). Section 165.845(2) reinforces the point by directing that "all persons in charge of law enforcement agencies and other criminal and juvenile justice agencies shall supply the department of justice with the information" requested.
- *Transfer to FBI/national compilation encouraged by law:* Though the crime data authorized by law "shall include data requested by the federal bureau of investigation under its system of uniform crime reports for the United States," there is no apparent direct mandate that the data be forwarded to the FBI.
- *State maintains own incident-based system; state program is NIBRS certified, with limited local agency participation to date:* As of the JRSA inventory, NIBRS-format jurisdictions covered 38 percent of Wisconsin's population and 45 percent of the state's crime. The state program received a NIBRS Planning grant.
- *Annual crime report not apparently referenced by law.*
- *Other key features:* In the same clause that mandates that local agencies report data to the state, the state Department of Justice is authorized to "conduct an audit to determine the accuracy of the data and other information it receives" from contributing agencies (§ 165.845(2)).
- *State Statistical Analysis Center:* Same entity, in the Wisconsin Department of Justice; indeed, the section of code authorizing UCR-type data collection directs that the department "maintain a statistical analysis center to serve as a clearing house of justice system data and information and conduct justice system research and data analysis under this section" (§ 165.845(1)(c)). The SAC is the Bureau of Justice Information and Analysis, in the department's Division of Law Enforcement Services.

WYOMING

- *Local law enforcement agencies required to submit crime statistics to state entity:* Wyo. Stat. § 7-1-102 vests authority in the state attorney general to "provide uniform forms for reporting all information necessary to obtain reliable statistics to ascertain the true condition of the crime situation in the state." Functionally, the authority is delegated to the Wyoming Division of Criminal Investigation. The statute obliges "all town, city, county and state law enforcement agencies, district courts, courts of limited jurisdiction, district attorneys, state adult and juvenile correctional institutions and state and local probation and parole agencies" to "maintain a public record of crime and criminals," and directs that they "shall furnish the information requested by the attorney general."

- *Transfer to FBI/national compilation not mentioned in law.*
- *State has no incident-based reporting system,* and was long labeled a state with no plan for NIBRS implementation. The Wyoming state UCR program did receive an NCS-X grant for basic planning of a a NIBRS-compliant system in 2015.
- *No apparent direct legal mandate for a crime report:* However, the Division of Criminal Investigation issued annual reports on crime in the state between 1977 and 2010, and instituted quarterly reports beginning in 2011.
- *State Statistical Analysis Center:* Sited separately from Uniform Crime Reporting Program, in Wyoming Survey and Analysis Center at the University of Wyoming.

– F –

Cautionary Tales from International Experience: Police-Report Crime Statistics in the United Kingdom

As described in Section 3.1, the United States represents an extreme case of decentralized authority both in terms of law enforcement and federal statistics—a combination with strong ramifications for collecting "national crime statistics." It is also a combination that sharply limits the degree to which the United States can directly learn from the experience of other countries in reengineering their crime statistics infrastructures. The circumstances are typically so varied in comparing the U.S. to other nations that the comparison is at best illustrative. Yet this report would be incomplete without mention of at least one key illustrative experience, simply as a check on overconfidence. We argue in this report for completely rethinking the way the United States measures crime, including adoption of a new classification of crime, and using that classification as blueprint for implementing and coordinating a new crime statistics system according to the principles and sensitivities of a statistical agency. We believe we have been appropriately open about the difficulty of the task. The story of another country that made these kinds of changes and still experienced a major crisis of confidence in its crime statistics is useful to review, strictly illustrative as it may be.

Similar to the United States, crime statistics in the United Kingdom historically drew from two principal sources: the British Crime Survey (BCS, akin to the U.S. National Crime Victimization Survey) and the collation of police-report data from about 45 territorial and special-purpose police agencies. However, for years, full authority for both crime data streams was vested in the same government ministry—the Home Office, responsible for administration of the police forces themselves. Over the years, several reviews of UK crime statistics were commissioned due to perceived diminishing of trust in the resulting data, much more so for the police-report (offense count) statistics than the BCS but still pervasive. As it turned out, these concerns over erosion of trust coincided with a major structural change in management of UK official statistics. The United Kingdom Statistics Authority was created in April 2008 pursuant to law enacted the previous year, and it was formed as an entity independent of any government ministry (accountable directly to Parliament). In addition to taking on oversight of the nation's statistical office (the Office for National Statistics, ONS), the Authority was also directed to certify that statistical series and products adhered to its Code of Practice for Official Statistics—and so to certify which statistical products would bear the label and designation as official "National Statistics" (and which would not).

As a major sector of official statistics, crime statistics quickly drew the Authority's attention. An initial report in 2010 summarized user input and previous reform suggestions—and ultimately sided with maintaining the organizational status quo (United Kingdom Statistics Authority, 2010:4–5). Specifically, the review noted that data collection for the BCS was already contracted out to a private survey organization, making transfer of authority from the Home Office to ONS relatively easy, but observed that it was unclear whether transfer of the police-report data to ONS would significantly improve trust in those data. Likewise, it concluded that "practical considerations"—the existing ties between the national Home Office and the territorial and special-purpose police agencies—advised against an administrative transfer of police-report data. An initial assessment of UK crime statistics effectively granted them conditional designation as "National Statistics" subject to completion of five enhancements, chief among them the provision of "more information about the quality of the police recorded crime statistics, including the consistency with which crimes are recorded and classified" (United Kingdom Statistics Authority, 2011:3).

However, by December 2010, the Home Office itself recognized the need for a more definitive direction and drew up terms of reference for a fuller Statistics Authority assessment. In addition to comprehensive review of the "gaps, discrepancies and discontinuities" in both the BCS and police-report data, the review was explicitly directed by the Home Office to "consider which body outside the Home Office is best placed to have future formal responsibility for the publication of crime statistics"—detaching the publication function from the compilation of the data—"and what should be the appropriate underpinning data collection arrangements." The National Statistician's review (United Kingdom Government Statistical Service, 2011) satisfied this request. It recommended that full authority for the BCS and that publication authority for all crime statistics be vested in the ONS and that collection/validation be formally detached from publication for police-report data. Effective April 2012, the primary recommendations were adopted: Authority for the renamed BCS, now the Crime Survey for England and Wales (CSEW), was transferred in full to ONS, while authority for police-report data was formally divided in function. The Home Office remained responsible for collecting police-report data, but was to channel compiled data to ONS—which was now a more public "owner" of the police-report data, analyzing and publishing them on a quarterly cycle. This elevated role of ONS in UK crime statistics was also the occasion for a reexamination of what constitutes a reportable offense in the data as well as the UK's intricate and voluminous set of "counting rules" for crimes in police-report data.

The National Statistician's review also called for assessment and strengthening of data quality and audit routines for police-report data, the regularly scheduled program of audits having been cancelled since 2007 due to resource constraints. Her Majesty's Inspectorate of Constabulary (2012) performed an independent review of recording practices by territorial police, updating an analysis performed in 2000; though it found improvement in the consistency of recording/reporting offenses, the audit suggested wider-than-desired variability in recording offenses (or "no criming"/not recording some offense types, such as sexual assault). Select committees of the House of Commons also heard allegations of misreporting, and particularly the "no criming" of incidents in order to meet performance targets and show reductions in crime levels. In January 2013, ONS researchers published an analysis of crime trends contrasting police-report and BCS/CSEW data, which suggested that the level of inconsistency of accurate police reporting might be increasing with time (Flatley and Bradley, 2013).

The transfer in authority for crime statistics publication and the ongoing concern over potential misreporting led the Statistics Authority to initiate another assessment of UK police-report statistics. This time, in January 2014, the consequences were severe. Specifically, the Statistics Authority stripped police-recorded crime data of "National Statistics" designation, citing "accumulating evidence [that] the underlying data on crimes recorded by the police may not be reliable" and that ONS (as publisher of the data) "does not have sufficient information about the quality of recorded crime data [to] ensure that users are made fully aware of the limitations of the recorded crime statistics" (United Kingdom Statistics Authority, 2014a:2). The Authority listed 16 specific requirements that ONS (working with the Home Office and other authorities) would have to satisfy to reacquire the National Statistics label. By and large, these requirements center around communication of uncertainty and data quality and the capacity of the resulting data to satisfy data user needs. For instance, the assessment said to "provide more information to users about [how] to interpret changes in the published crime statistics from one reference period to another," "explain more fully how all the administrative data sources are used to produce statistics about crime," and (mindful of ongoing audit by the Inspectorate of Constabulary) consider the general question of "whether statistics based on police recorded crime data can be produced to a level of quality that meets users' needs" (United Kingdom Statistics Authority, 2014a:4). Another requirement is to make public information "about police forces' progress in switching to use of the Data Hub"—a Home Office–created system by which local departments transfer their crime data directly to the Home Office for compilation—"and the reason why this has taken so long" (United Kingdom Statistics Authority, 2014a:5).

Subsequent to the stripping of National Statistics status, a new Crime Statistics Advisory Committee was empaneled to assist in the process of working through the requirements. The House of Commons' Public Administration Select Committee (2014:3, 22) suggested approval of the Authority's conclusions but concluded that the Authority, the ONS, and the Home Office had all been "far too passive in the face of concerns raised about [Police Recorded Crime (PRC)]; they have repeatedly missed opportunities to ensure the integrity and quality of PRC data." In particular, the committee chided the Authority for having initially designated police-report data as National Statistics in 2011, and found it "deplorable that ONS can have overseen the production of crime statistics [with] what appears to have been very limited knowledge of the 'quality assurance' steps" performed on the data. Further, the committee recommended "the re-instatement of annual audits of crime recording practices," noting that their suspension in 2007 was a "mistake," and "deprecate[d] the use of targets" for reduction in crime tallies by local departments "in the strongest possible terms" as a detriment to data integrity.

Not long after the Statistics Authority stripped police-report data for England and Wales of its National Statistics designation, it did the same for police-report data for Scotland (United Kingdom Statistics Authority, 2014b). Two years later, the Scottish government was notified that its *Recorded Crime in Scotland* data were reinstated as National Statistics (Humpherson, 2016). As of mid-2017, however, the police-report data for England and Wales continue to be decertified as National Statistics.

One could look at the division of labor established in 2012 in the UK—substituting "FBI" for "Home Office" and "BJS" for "ONS"—and attempt some inference concerning our delineation of coordination and governance rules for crime statistics in Chapter 3. This is not at all what we intend by telling this story, and we do not support such easy inference; much like blithely comparing community crime levels absent any context, there are sufficient differences between the United States and United Kingdom crime measurement circumstances that simple comparison is misleading. But some broader take-away lessons are justified, from the general nature of the UK experience and from the substance of the UK Statistics Authority's rationale for decertification of police-report data. To wit, the UK experience and the example of Ireland (discussed in Section 4.1.4 of Report 1)[1] support the conclusion that adoption of a rigorous classification and implementation of new systems are not panaceas on their own. Indeed, they are largely for naught without adequate attention to the quality of the underlying data entering the system or to whether the resulting data products are understood by data users/stakeholders and meet their needs.

[1]In brief: Ireland was one of the first countries (in recent years) to fully revisit and revise its classification of crime; its result was an important starting point for the United Nations Office on Drugs and Crime (UNODC) International Classification of Crime for Statistical Purpose (ICCS) on which our proposed classification is patterned. Implementation of the new classification was concomitant with transfer of responsibility for crime statistics publication from the national police (La Garda Síochána) to the Irish Central Statistical Office (CSO). However, a review found "systemic failures in recording practices" in the Police Using Leading Systems Effectively (PULSE) system—basically, Ireland's UCR database, the recording system used to compile inputs from local police offices (Garda Síochána Inspectorate, 2014:9–11). Misclassification results were sufficiently large that the CSO formally suspended the publication of Irish police-report crime statistics for 6 months, only resuming their publication after completing their own review (Central Statistics Office, 2015) and instituting remedies and data improvements.

– G –

Biographical Sketches of Panel Members and Staff

Janet L. Lauritsen *(chair)* is Curators' Distinguished professor of criminology and criminal justice at the University of Missouri–St. Louis. Her research focuses on the causes and consequences of victimization, the social and historical contexts of crime and victimization, and quantitative research methodologies. Her most recent publications cover topics such as racial and ethnic disparities in risk for violence during the past four decades, gender inequality and violence against women, repeat victimization, and the relationship between changing economic conditions and violent victimization. Her research has been funded by the National Science Foundation, the Bureau of Justice Statistics, and the National Institute of Justice. She currently serves as co-editor of the journal *Criminology*. At the National Academies of Sciences, Engineering, and Medicine, she has previously served as a member of the Committee on Law and Justice, the Roundtable on Crime Trends, the Panel to Review the Programs of the Bureau of Justice Statistics, and the Committee to Review Research on Police Policies and Practices. She has a Ph.D. in sociology from the University of Illinois at Urbana-Champaign.

Daniel B. Bibel retired in February 2015 as program manager of the Crime Reporting Unit of the Massachusetts State Police since 1992. Previously, he served as program manager of the Uniform Crime Reporting program for the Commonwealth of Massachusetts Criminal History Systems Board and as director of the Massachusetts Committee on Criminal Justice's Statistical Analysis Center. He has held adjunct faculty positions at the Northeastern

University School of Criminal Justice and at the University of Massachusetts, Lowell. A two-time past president of the Association of State Uniform Crime Reporting Systems, he is currently a member of the Uniform Crime Reporting Subcommittee of the Federal Bureau of Investigation Advisory Policy Board, and served for nearly a decade on the Statewide Geographic Information Advisory Committee in Massachusetts. He holds a bachelor's degree in sociology from Fairleigh Dickinson University and an M.S. degree in criminal justice from Northeastern University.

Jonathan P. Caulkins is H. Guyford Stever professor of operations research and public policy at the Heinz College of Information Systems and Public Policy, Carnegie Mellon University. A member of the Carnegie Mellon faculty since 1990, he has taken leaves of absence to teach at the university's campus in Doha, Qatar, to direct RAND's Drug Policy Research Center, and to establish RAND's Pittsburgh office. His research focuses on mathematical models of social policy problems and on specific policies in the area of drugs, crime, and delinquency. He has consulted with a wide variety of agencies and organizations, including the Central Intelligence Agency, the United Nations Office on Drugs and Crime, the United Kingdom Drug Policy Commission, and Westat. He was elected to the National Academy of Engineering in 2015. He is a fellow of the Institute for Operations Research and the Management Sciences and won the David Kershaw Award from the Association of Public Policy Analysis and Management. At the National Academies, he has served on the Committee on Increased Costs to the U.S. Department of Justice of Border Enforcement by the U.S. Department of Homeland Security, the Committee on Reducing Tobacco Use, and the Committee on Immunotherapies for Treating Drug Addiction. He holds a bachelor's degree in systems science and engineering, computer science, and engineering and policy and an M.S. degree in systems science and engineering, both from Washington University, and an S.M. degree in electrical engineering and computer science and a Ph.D. degree in operations research from the Massachusetts Institute of Technology.

Daniel L. Cork *(study director)* is a senior program officer for the Committee on National Statistics (CNSTAT), currently serving as study director for this panel and for the Standing Committee on Reengineering Census Methods. He joined the CNSTAT staff in 2000 and has served as study director or program officer for several census-related panels, including the Panels on Residence Rules in the Decennial Census, Research on Future Census Methods (2010 Planning panel), Review of the 2000 Census, and Review of the 2010 Census. He also directed the Panel to Review the Programs of the Bureau of Justice Statistics (in cooperation with the Committee on Law and Justice) and was senior program officer for the Panel on the Feasibility, Accuracy, and Technical Capability of a National Ballistics Database (joint with the Committee on Law and Justice and the National Materials Advisory Board). His research interests include quantitative criminology, geographical analysis, Bayesian statistics, and statistics in sports. He has a B.S. in statistics from George Washington University and an M.S. in statistics and a joint Ph.D. in statistics and public policy from Carnegie Mellon University.

Kim English is research director for the Division of Criminal Justice, within the Colorado Department of Public Safety, and director of the Colorado Statistical Analysis Center. In this capacity, she manages a staff of professional researchers engaged in a variety of criminal and juvenile justice research and policy analysis activities, providing research support, policy analysis, and program evaluation findings to the Colorado General Assembly, the governor's office and the Colorado Criminal and Juvenile Justice Commission. She received the 1999 J. Paul Sylvestre Award from the Justice Research and Statistics Association and the Bureau of Justice Statistics for outstanding achievement in advancing criminal justice statistics in the states. She has consulted with the Council of State Governments, the National Governor's Association, the National Association of State Legislatures, the National Institute of Corrections, the Center for Sex Offender Management, the American Probation and Parole Association, and criminal justice agencies in states across the country. She has an M.A. degree in sociology and criminology from the University of Colorado at Boulder.

Robert M. Goerge is a senior research fellow at Chapin Hall at the University of Chicago, where he also serves as director of the Joint Program on Public Policy and Computing at the Harris School for Public Policy Studies. His research focuses on improving available data on children and families, in particular those affected by violence or who require special services related to disability or poverty. He developed Chapin Hall's Integrated Database on Child and Family Programs in Illinois, which combines data from educational, criminal/juvenile justice, employment, and healthcare programs with administrative data on receipt of social services. With NORC at the University of Chicago, he is also principal investigator of the National Survey of Early Care and Education. At the National Academies, he served on the Panel on Data and Methods for Measuring the Effects of Changes in Social Welfare Programs. He has a Ph.D. in social policy from the University of Chicago.

Nola M. Joyce retired in early 2016 after serving since February 2008 as deputy commissioner of the Philadelphia Police Department. In that capacity, she headed the Office of Organizational Services, Strategy, and Innovations and led the departmental change management efforts for Commissioner Charles Ramsey. From 1998 to 2007, she performed similar duties with the Metropolitan Police Department, Washington, D.C.; previously, she spent 6 years as deputy director of the Chicago Police Department's Research and Development Division, where she was particularly involved in implementing the Chicago Alternative Policing Strategy model for community policing. For 11 years, she was manager of the research, planning, and budget divisions of the Illinois Department of Corrections. She received the Gary P. Hayes award in 2010 from the Police Executive Research Forum; she is a member of the International Association of Chiefs of Police Research Advisory Committee and the U.S. Department of Justice, Bureau of Justice Assistance, Law Enforcement Futurist Group. She has two master's degrees from Southern Illinois University—one in urban affairs and public policy and another in sociology with specialization in research methodology and statistics—and another master's degree in homeland defense and security from the U.S. Naval Postgraduate School; she is also completing work in Temple University's Ph.D. program in criminal justice.

James P. Lynch *(consultant to the panel)* is professor and chair of the Department of Criminology and Criminal Justice at the University of Maryland. From June 2010 through his appointment at the University of Maryland, he served as director of the Bureau of Justice Statistics. Previously, he was distinguished professor at John Jay College of Criminal Justice in New York. At the Bureau of Social Science Research in the 1980s, he served as manager of the National Crime Survey redesign effort for the bureau. He became a faculty member in the Department of Justice, Law, and Society at American University in 1986, where he remained as associate professor, full professor, and chair of the department until leaving for John Jay in 2005. He has published 3 books, 25 refereed articles, and over 40 book chapters and other publications. He was elected to the executive board of the American Society of Criminology in 2002 and has served on the editorial boards of *Criminology* and the *Journal of Quantitative Criminology* and as deputy editor of *Justice Quarterly*. He has also chaired the American Statistical Association's Committee on Law and Justice Statistics. He was a member of the Panel to Review the Programs of the Bureau of Justice Statistics, and currently serves on the Committee on Law and Justice and the Panel on Improving Federal Statistics for Policy and Social Science Research Using Multiple Data Sources and State-of-the-Art Estimation Methods. He has a Ph.D. in sociology from the University of Chicago.

David McDowall is distinguished teaching professor of criminal justice at the University at Albany, State University of New York, and co-director of the Violence Research Group. Prior to joining the Albany faculty, he was a postdoctoral fellow at the University of Michigan and served on the faculties of the University at Buffalo and the University of Maryland. His research makes extensive use of quantitative methods and aggregate crime data. He has published on policy issues in juvenile justice and firearm violence, and on the nature and statistical properties of crime rate trends. From 2001–2008, he was editor-in-chief of the *Journal of Quantitative Criminology*. He currently serves on the editorial boards of the *Journal of Experimental Criminology*, the *Journal of Quantitative Criminology*, and the *Journal of Research on Crime and Delinquency*. A fellow of the American Society of Criminology, he received the Criminology Teaching Award for career-length achievement from the society in 2011. He has a Ph.D. in sociology from Northwestern University.

Jennifer H. Madans was named associate director for science of the National Center for Health Statistics (NCHS), U.S. Centers for Disease Control and Prevention, in May 1996, and is responsible for the overall plan and development of NCHS's data collection and analysis programs. Her research concentrations include data collection methodology, measurement of health and functioning, and health services research; she was founding member and chair of the steering committees for three United Nations-sponsored initiatives to develop internationally comparable measures of disability and health. She has directed two major national longitudinal studies (the National Health and Nutrition Examination Survey I Epidemiologic Followup Study and the National Nursing Home Followup Study), and participated in the redesign of the National Health Interview Survey questionnaire. She has also served as adjunct associate professor in the Demography and Community and Family Medicine Departments at Georgetown University. She is a fellow of the American Statistical Association, an elected member of the International Statistical Institute, and a past vice president of the International Association of Official Statistics. At the National Academies, she is currently a member of the Panel on Addressing Priority Technical Issues for the Next Decade of the American Community Survey; she previously served on the Committee to Develop Criteria for Evaluating the Outcomes of Approaches to Prevent and Treat Obesity and the steering committee for the Workshop on the Future of Federal Household Surveys. She has a B.A. degree from Bard College and M.A. and Ph.D. degrees in sociology from the University of Michigan.

Michael D. Maltz is professor emeritus of criminology, law, and justice and of information and decision sciences at the University of Illinois at Chicago and senior research scientist and adjunct professor of sociology at the Criminal Justice Research Center at Ohio State University. A past editor of the *Journal of Quantitative Criminology*, he was a visiting fellow at the Bureau of Justice Statistics from 1995 to 2000 and was formerly a staff member of the National Institute of Law Enforcement and Criminal Justice (now the National Institute of Justice). His primary research interests have been on the validity of crime data, including routines for handling of missing data, and the capability for drawing useful inferences from those data through mathematical models and visualization techniques. His 1984 book *Recidivism* was awarded the Lanchester Prize of the Operations Research Society of America (later, the Institute for Operations Research and the Management Sciences). In 1996, he held a Fulbright Scholarship at El Colegio de Michoacán in Mexico. He holds a Ph.D. in electrical engineering from Stanford University.

Michael C. Miller has over 24 years of law enforcement experience where he has worked at the city, county, tribal, and federal levels. Currently, he is serving as assistant chief of police, administrative services, for the Coral Gables Police Department, Florida. Prior to joining the Coral Gables Police Department in early 2013 as administrative services director, he served in the Federal Bureau of Investigation (FBI) for 6 years as the special assistant to the executive assistant director of the Criminal, Cyber, Response and Services Branch and as the special assistant to the assistant director of the Directorate of Intelligence. He also served for 2 years as an FBI detailee to the U.S. Bureau of Indian Affairs as deputy associate director of law enforcement operations. Prior to joining the FBI, he spent 13 years as a management consulting executive where he held several increasingly responsible positions focusing on public safety and law enforcement. Most notably, he served for 3 years as the global program executive for Accenture's Immigration, Justice and Public Safety practice. He also spent 11 years as regular and reserve deputy sheriff in Wise County, Texas, and 5 years as a reserve police officer in Addison, Texas. He holds a B.S. degree in biomedical science from Texas A&M University and an M.P.A. from Harvard University's John F. Kennedy School of Government, and has completed the executive program in Navigating Strategic Change at Northwestern University's Kellogg School of Management.

James J. Nolan, III, is professor of sociology and criminology at West Virginia University where he teaches courses about crime and social control. His research currently focuses on police procedures, crime measurement, and hate crime. From 1980 to 1993, he worked in the Wilmington, Delaware, police department, starting as patrol officer and eventually serving as project director for the community-oriented policing part of the Weed and Seed Program, for which Wilmington was one of the original 16 sites. He also served as senior policy advisor to the Secretary of Public Safety in Delaware. He graduated from the FBI National Academy in 1992 and was chief of the Crime Analysis, Research, and Development Unit in the FBI's Criminal Justice Information Services Division (CJIS) from 1995 to 2000, at which time he joined the West Virginia University faculty. During his tenure at CJIS, he provided management oversight for the National Hate Crime Data Collection Program. In 2010, he was named West Virginia's Professor of the Year by the Carnegie Foundation for the Advancement of Teaching and the Council for

Advancement and Support of Education. He is a past recipient of a grant to study statistical adjustments to police crime data from the American Statistical Association and the Bureau of Justice Statistics. He holds B.S. and M.S. degrees from Wilmington College and a M.Ed. and a Ph.D. in psychoeducational processes from Temple University.

Amy O'Hara joined the Stanford Institute for Economic Policy Research in July 2017. From 2014 until her move to Stanford, she was chief of the Center for Administrative Records Research and Applications (CARRA), which was then part of the research and methodology directorate at the U.S. Census Bureau. She began her career at the Census Bureau in 2004 as an economist/statistician in the Social, Economic, and Housing Statistics Division before shifting to CARRA in 2008. Among other accomplishments in attempting to integrate administrative records data into the full suite of Census Bureau processes, she led the 2010 Census Match Study—an unprecedented complete match/linkage of the full set of returns from the 2010 decennial census to a composite of administrative records data from 8 federal agencies, used to examine differences in coverage and possible sources of error in both sources. She received an Arthur S. Flemming Award, recognizing outstanding achievement and leadership in federal government service, from the Trachtenberg School of Public Policy and Public Administration, George Washington University, in 2012. She has M.A. and Ph.D. degrees in economics from the University of Notre Dame.

John V. Pepper is professor of economics at the University of Virginia, where he has served on the faculty since 1996. His work examines identification problems that arise when evaluating a wide range of public policy questions including such subjects as health and disability programs, welfare policies (e.g., SNAP), and drug and crime policies. From 2001 to 2002, he served as study director of the National Academies' Committee to Improve Research Information and Data on Firearms, culminating in the 2004 report *Firearms and Violence: A Critical Review*. He was a member of the Committee on Improving Evaluation of Anti-Crime Programs and served as consultant to other panels of the National Academies' Committee on Law and Justice, most recently serving as co-editor with Daniel S. Nagin of *Deterrence and the Death Penalty* (2012), the report of a study panel of the same name. He holds a B.A. degree in quantitative economics from Tufts University and M.S. and Ph.D. degrees in economics from the University of Wisconsin–Madison.

Alex R. Piquero is Ashbel Smith professor of criminology and associate dean for graduate programs in the School of Economic, Political, and Policy Sciences at the University of Texas at Dallas. He is also adjunct professor at the Key Centre for Ethics, Law, Justice, and Governance, Griffith University, Life

Course Centre Fellow at the University of Queensland, and faculty affiliate, Center for Violence and Injury Prevention, George Warren Brown School of Social Work, Washington University in St. Louis. From 2008–2013 he was co-editor of the *Journal of Quantitative Criminology*. Prior to arriving at UT-Dallas, he was on the faculties of Florida State University, University of Maryland, John Jay College of Criminal Justice/City University of New York, University of Florida, Northeastern University, and Temple University. He is a fellow of both the American Society of Criminology and the Academy of Criminal Justice Sciences and in 2014 was awarded The University of Texas Board of Regents' Outstanding Teaching Award. At the National Academies, he served on the Committee on Assessing the Research Program of the National Institute of Justice as well as The Panel on a Prioritized Plan to Implement a Developmental Approach in Juvenile Justice Reform. He holds B.A., M.A., and Ph.D. degrees in criminology from the University of Maryland.

Jeffrey L. Sedgwick is executive director of the Justice Research and Statistics Association in Washington, DC, the support and coordination arm of the Bureau of Justice Statistics–funded state Statistical Analysis Centers. Previously, he was co-founder and managing partner of Keswick Advisors, a statistical analysis and policy evaluation consulting firm in Richmond, Virginia. He was confirmed by the Senate as director of the Bureau of Justice Statistics (BJS) in March 2006. In January 2008, he was named acting assistant attorney general for the Office of Justice Programs (OJP) in addition to the BJS directorship; subsequently nominated in April as assistant attorney general for OJP, he was confirmed to the post in October 2008 and served in that capacity until the change in presidential administrations in January 2009. He has previously served from January 1984 to January 1985 as deputy director for data analysis at BJS. He is professor emeritus of political science at the University of Massachusetts Amherst, where he joined the faculty in 1977 and where he specialized in aspects of American government including public finance, criminal justice policy, and the role of the American presidency. He has also served as visiting instructor of leadership studies at the University of Richmond. He is the author of *Law Enforcement Planning: The Limits of an Economic Approach* (1984) and *Deterring Criminals: Policymaking and the American Political Tradition* (1980), and has directed or participated in several international programs, teaching and lecturing in Ukraine, Kyrgyzstan, Germany, Trinidad and Tobago, and other countries. He has a bachelor's degree in political science from Kenyon College and a master's degree in public administration and public policy and Ph.D. in government and public affairs from the University of Virginia.

Paul K. Wormeli *(liaison member, Committee on Law and Justice, and consultant to the panel)* is principal and innovation strategist at Wormeli Consulting, LLC, a technology consulting firm for public service agencies. He is executive director emeritus of the Integrated Justice Information Systems Institute, a nonprofit corporation in Ashburn, Virginia, formed to help state and local governments develop ways to share information among the disciplines engaged in law enforcement and the administration of justice. He has had a long career in the field of law enforcement and justice technology. He has been active in the development of software products, has managed system implementation for dozens of agencies throughout the world, and has managed national programs in support of law enforcement and criminal justice agencies. He was the first national project director of Project SEARCH, and was subsequently appointed as deputy administrator of the Law Enforcement Assistance Administration in the U.S. Department of Justice (a predecessor of the current Office of Justice Programs). He helped design the first mobile computing equipment sold in this country to law enforcement agencies. He managed the staff work and wrote much of the report for the Information Systems section in the report of the National Commission on Standards and Goals for Criminal Justice, which dealt with criminal justice information system standards. He has been an advisor to the White House on security and privacy, participated in the drafting of federal law on this topic, and was responsible for the development of numerous state plans to implement the federal and state laws on information system security and privacy. During his tenure in the Justice Department, he served on the President's Committee on Drug Enforcement. He was also the first chairman of the Integrated Justice Information Systems Industry Working Group (IWG), a consortium of over 100 companies that was formed at the request of the U.S. Department of Justice to help facilitate the implementation of Integrated Justice Information Systems throughout the nation. At the National Academies, he is a current member of the Committee on Law and Justice. He holds a B.S. degree in electronics engineering from the University of New Mexico and an M.S. degree in engineering administration from George Washington University.

COMMITTEE ON NATIONAL STATISTICS

The Committee on National Statistics was established in 1972 at the National Academies of Sciences, Engineering, and Medicine to improve the statistical methods and information on which public policy decisions are based. The committee carries out studies, workshops, and other activities to foster better measures and fuller understanding of the economy, the environment, public health, crime, education, immigration, poverty, welfare, and other public policy issues. It also evaluates ongoing statistical programs and tracks the statistical policy and coordinating activities of the federal government, serving a unique role at the intersection of statistics and public policy. The committee's work is supported by a consortium of federal agencies through a National Science Foundation grant, a National Agricultural Statistics Service cooperative agreement, and several individual contracts.